T0245157

Applied ethology 2015:

Ethology for sustainable society

ISAE2015

Proceedings of the 49th Congress of the International Society for Applied Ethology

14-17 September 2015, Sapporo Hokkaido, Japan

Ethology for sustainable society

edited by:

Takeshi Yasue

Shuichi Ito

Shigeru Ninomiya

Katsuji Uetake

Shigeru Morita

Wageningen Academic
P u b l i s h e r s

EAN: 9789086862719
e-EAN: 9789086868179
ISBN: 978-90-8686-271-9
e-ISBN: 978-90-8686-817-9
DOI: 10.3920/978-90-8686-817-9

First published, 2015

© Wageningen Academic Publishers
The Netherlands, 2015

The individual contributions in this publication and any liabilities arising from them remain the responsibility of the authors.

The publisher is not responsible for possible damages, which could be a result of content derived from this publication.

'Sustainability for animals, human life and the Earth'

On behalf of the Organizing Committee of the 49th Congress of ISAE 2015, I would like to say fully welcome for all of you attendances to come to this Congress at Hokkaido, Japan!

Now a day, our animals, that is, domestic, laboratory, zoo, companion, pest and captive animals or managed wild animals, and our life are facing to lots of problems. The one of the most important problems should be sustainability of our society. That's meaning a sustainability of the Earth. Thus, the main theme of this Congress that we set is 'Ethology for Sustainable Society'. For this main theme, our viewpoints are not only for welfare of farm animals and its practices but also for consideration of captive animals and wild animals as well as human-animal interactions. Japan has been traditionally 'rice country', though we are also keeping 3,860,000 cattle of dairy and beef, 9,540,000 swine and 308,000,000 chickens, approximately. On the other hand, a population of our farmers has been declined rapidly in this decade. Additionally, our agricultural damage from Sika deer, wild boars, Japanese monkeys, and pest birds would be over 20 billion Japanese yen (about US$ 160,000,000) per year. These problems could occur for everywhere in the world. In such situation, we should consider the sustainability of these animals and human as a society from the ethological viewpoint.

Hokkaido is the northernmost island in Japan, and old and new. Up to 150 years ago, AINU people had been mainly living this island as hunter-gatherer. Thereafter, rapid development has been occurred, so this island now is the greatest agricultural area in Japan. Sapporo is the capital city of Hokkaido, having about 2 million populations. She has been famous international city since winter Olympic at 1972. Her summer is cool and dry, but heavy snow in winter. Although Sapporo is now so big city, wild brown bears sometime visit south and west area of this city. Toyohira River, that runs a centre of the city, salmon go back in autumn. If you lucky, you can see. Hokkaido University, is a Congress place, is one of the oldest universities in Japan, established as a Sapporo Agricultural College since 1876. You can see several historic buildings including barns in the campus.

Enjoy our Ethology Congress with fruitful discussion and Sapporo life!.

Seiji Kondo
Hokkaido University

Acknowledgements

Organizing committee

President
Seiji Kondo — Hokkaido University

Advisory committee

Shusuke Sato	Teikyo University of Science
Toshio Tanaka	Azabu University
Shinji Hoshiba	Rakuno Gakuen University
Ryo Kusunose	Japan Farriery Association
Fumiro Kashiwamura	Obihiro University of Agriculture and Veterinary Medicine
Shizufumi Ebihara	Kwansei Gakuin University
Naoshige Abe	Tamagawa University

Executive committee

Local acting committee

Chair
Shigeru Morita — Rakuno Gakuen University

Vice-chair

Hiroshi Yamada	Rakuno Gakuen University
Koichiro Ueda	Hokkaido University
Hiroki Nakatsuji	Rakuno Gakuen University
Masahito Kawai	Hokkaido University
Tetsuya Seo	Obihiro University of Agriculture and Veterinary Medicine
Kenichi Izumi	Rakuno Gakuen University
Keiko Furumura	Obihiro University of Agriculture and Veterinary Medicine
Keiji Takahashi	Rakuno Gakuen University
Tomohiro Mitani	Hokkaido University
Masato Yayota	Gifu University

Information committee

Chair

Takeshi Yasue   Ibaraki University

Vice-chair

Masato Aoyama   Utsunomiya University

Shuichi Ito   Tokai University
Michiru Fukasawa   NARO/TARC

Endowment fund management

Ken-ichi Yayou   National Institute of Agrobiological Sciences

Treasurer

Yoshitaka Deguchi   Iwate University
Akihiro Matsuura   Kitasato University

Scientific Committee

Chair

Katsuji Uetake   Azabu University

Vice-chair

Shigeru Ninomiya   Gifu University

Ken-ichi Takeda   Shinshu University
Daisuke Kohari   Ibaraki University
Masaki Tomonaga   Kyoto University
Yoshitaka Nakanishi   Kagoshima University
Nobumi Hasegawa   University of Miyazaki
Akihisa Yamada   NARO/KARC
Takami Kosako   NARO/NILGS
Yoshie Kakuma   Teikyo University of Science
Megumi Fukuzawa   Nihon University
Yusuke Eguchi   NARO/WARC
Hideharu Tsukada   Azabu University
Yoshikazu Ueno   Higashiyama Zoo & Botanical Garden
Eiichi Izawa   Keio University
Shingo Tada   NARO/HARC
Tsuyoshi Shimmura   National Institute for Basic Biology
Tadatoshi Ogura   Kitasato University
Yumi Yamanashi   Kyoto University
Mizuna Ogino   Ishikawa Prefectural University
Chihiro Kase   Chiba Institute of Science

Reviewers

Appleby, M.C.
Blokhuis, H.J.
Buijs, S.
De Passille, A.M.
Eguchi, Y.
Fukuzawa, M.
Harris, M.
Hasegawa, N.
Izawa, E.
Jensen, M.B.
Kakuma, Y.
Kase, C.
Keeling, L.
Koene, P.
Kohari, D.
Makagon, M.
Mänd, M.
Marchant-Forde, J.
Nakanishi, Y.
Newberry, R.C.
Nielsen, B.L.

Ninomiya, S.
Paranhos Da Costa, M.
Ogino, M.
Ogura, T.
Rault, JL.
Rushen, J.
Rutter, S.M.
Shimmura, T.
Spoolder, H.
Takeda, K.
Tomonaga, M.
Tucker, C.
Tsukada, H.
Ueno, Y.
Uetake, K.
Valros, A.
Veissier, I.
Von Borell, E.
Weary D.
Yamada, A.
Yamanashi, Y.

Sponsors

Gold sponsors

Silver sponsors

住友化学園芸株式会社

Dr. Shimojo, M. (private donation)

Purina is Proud to Sponsor
The 49th International Congress of the International Society for Applied Ethology

競 走 馬 用 飼 料 ・ 資 材 の パ イ オ ニ ア

競走馬用飼料

エン麦、輸入牧草、配合飼料
添加物飼料

牧場用資材

牧柵、肥料、麦稈、稲藁
木質系敷料

ファーム
コンサルティング

株式会社ホクチク 北海道浦河郡浦河町西幌別328-66

http://www.hokuchiku.com

発情発見
だけじゃない

健康管理と
正確な発情発見

高い受胎率は、個体の健康状態と同様に、個体乳量の増加に大きく関与します。よって正確な発情発見とともに、個々の健康状態を自動的に確認できることは、農場経営にとても有益なことです。
ネダップのヘルスモニタリング＆発情発見システムは、24時間各個体の行動と活動を監視し、発情している個体と健康状態に疑いのある個体を発見します。

システムは自動的に、発情している個体と健康状態に特別な注意が必要な個体に関する正確な情報を提供します。これらのデータにより、適切な時期の授精を確実に行い、健康状態の異常を早い段階で発見することが可能です。

ネダップのヘルスモニタリング＆発情発見システムは、各個体のベストなパフォーマンスを引き出し継続させる良きパートナーとなります。

Nedap Heat Detection

| 24時間 正確な発情発見 | 90%以上の 発情発見率 | 半径150m以上の アンテナ検知範囲 | 健康状態の 常時監視 | 首タイプと 脚タイプで 利用可能 | ISO 個体識別機能を 利用可能 | 適切な管理による 個体の生産量の 大きな向上 |

technology that matters

株式会社 土谷特殊農機具製作所
www.tsuchiyanoki.com

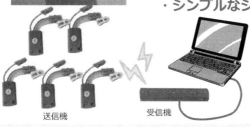
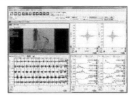
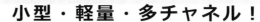

きみといきる。
ふくしまプロジェクト
FUKUSHIMA PROJECT

http://fukushimaproject.org

Sapporo

Sapporo is the capital of Hokkaido and Japan's fifth largest city. In 1857, the city's population stood at just seven people. Sapporo became world famous in 1972 when the Olympic Winter Games were held there. Today, the city is well known for 'Ramen', 'beer', and 'the annual snow festival' held in February.

City Information

http://www.welcome.city.sapporo.jp/
http://www.conventionsapporo.jp/

Sapporo Info

'Sapporo Info' is available in English, Chinese, Korean and Thai apart from Japanese. This application provides offline caching. People from other countries can also enjoy tourist maps using this application.

WiFi system

The free WiFi service is provided by some carriers in Japan. For example, http://flets.com/freewifi/index.html is operated by the NTT East Co.
In the Conference Hall building, registered guests can use a free-WiFi service based on the 'eduroam' system on campus. We give you the ID and password to use for authentication at the registration desk.

Venue

Hokkaido University

The Congress will be held in Hokkaido University. Once arriving at Sapporo station, the main gate of the campus is a short and flat walk.

Hokkaido University Conference Hall

From the Sapporo station to the congress venue in Hokkaido University, you can walk within 15 minutes, 2 blocks north (200m) and 1 blocks west (100m).

Official language of meeting is English.
Your name badge should be worn at all times at the conference and at conference dinner.
We give Lunch-guide near the venue. The lunch is not included in registration fee.

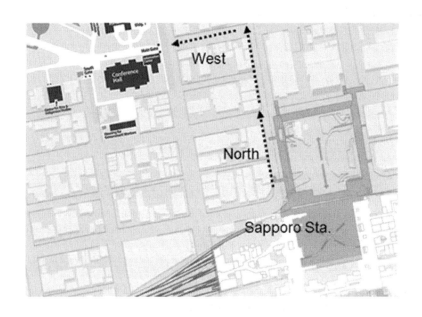

West

North

Sapporo Sta.

1st Floor

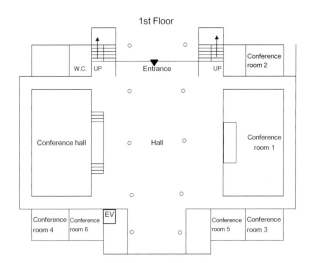

2nd Floor

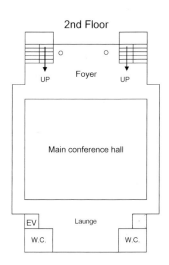

Scientific and Social Program

Monday 14 September

	Main conference hall	Conference hall
08:30	Morning Welcome Coffee receprion	
10:00	Opening Ceremony	
10:30-11:30	**Wood-Gush Memorial Lecture** **A study on the normal behaviour of cattle in Japan. Shusuke Sato (p. 39)**	
11:30-12:15	**PLENARY: Applied ethology for ever: animal management and welfare are integral to sustainability. Michael C Appleby (p. 40)**	
12:15	Lunch	
14:00-14:45	**PLENARY: Animal welfare assessment: the US perspective. Cassandra Tucker (p. 41)**	
	SESSION 1: Animal welfare assessment for good farm practice and production	SESSION 2: Free Topics
14:50	Producer estimations and perceived importance of lameness in dairy cattle: where do we move from here? (p. 45)	Startling pigs - a new measure of pig affect and welfare (p. 53)
15:05	The influence of gentle interactions during milking on avoidance distance, milk quality and milk flow in dairy cows (p. 46)	Repeated testing in a cognitive bias paradigm in pigs (p. 54)
15:20	Human-animal interactions and safety during dairy cattle handling (p. 47)	Is the Cognitive bias test able to assess changes in emotional state due to housing conditions in pigs? (p. 55)
15:35	Daily lying behaviour as an indicator of pregnancy toxemia in dairy goats (p. 48)	
15:50	Coffee Break	
16:20	What can carcass-based assessments tell us about the lifetime welfare of pigs? (p. 49)	The association between range use as monitored by RFID, fearfulness and plumage condition in free-range laying hens (p. 56)
16:35	From farm to slaughter: broiler welfare assessment during catching, loading, transport and lairage (p. 50)	Auto-detecting cattle behaviour using IR depth-sensor camera (p. 57)
16:50	Relationship between lameness scored on farm and foot pad lesions scored at the abattoir in broiler chicken flocks (p. 51)	Evaluating health of jungle crow using animal-borne accelerometers (p. 58)
17:05	Reliability of the visual gait score as a method of gait assessment in domestic ducks (p. 52)	
18:00 - 20:30	Workshop	

Tuesday 15 September

	Main conference hall	Conference hall
09:00- 09:45	**PLENARY: How growing environment effects on reproductive behavior? – taking Orangutan (genus *Pongo*) for example. Noko Kuze (p. 42)**	
	SESSION 3: Free Topics	SESSION 4: Free Topics
09:50	Relations between play, vocalization and energy intake of dairy calves following separation from the cow (p. 59)	The effects of group lactation sow housing on aggression and injuries in weaner pigs. (p. 63)
10:05	Weaning calves off milk according to their ability to eat solid feed reduces hunger and improves weight gain during weaning. (p. 60)	Investigating Sow Posture Changes: Does Environment Play a Role? (p. 64)
10:20	Effect of two-stage weaning on behaviour and lying time in horses (p. 61)	Effects of floor space allowance on aggression and stress in grouped sows (p. 65)
10:35	Effects of maternal vocalisations on the domestic chick stress response (p. 62)	Lameness and claw lesions in group housed sows: effect of rubber-topped floors (p. 66)
10:50	Cofee Break	
11:10	Short time oral presentations (poster Group 1)	
12:00	Lunch	
13:30	Poster session Group 1 (p. 101-156)	
	SESSION 5: Freedom to express normal behaviour in captive animals	SESSION 6: Free Topics
14:30	Promoting positive welfare through environmental enrichment (p. 67)	Ammonia on a live export shipment, effects on sheep behaviour and development of an effective ammonia sampling strategy (p. 74)
14:45	Environmental enrichment in kennelled Pit Bull Terriers (*Canis lupus familiaris*) (p. 68)	Effects of bedding with recycled sand on lying behaviours of late-lactation Holstein dairy cows (p. 75)
15:00	Do farmed American mink (*Neovison vison*) prefer to eat using more naturalistic postures? (p. 69)	Cooling dairy cows efficiently with water: Effects of sprinkler flow rate on behavior and body temperature (p. 76)
15:15	Housing implications on the behavioural development of red fox (*Vulpes vulpes*) cubs (p. 70)	Breeding but not management change may simultaneously reduce pig aggressiveness at regrouping and in stable social groups (p. 77)
15:30	Cofee Break	
15:45	Normal behaviour of housed dairy cattle with and without pasture access: a review (p. 71)	Effect of Social Stressors on Behaviour and Faecal Glucocorticoid in Pregnant Ewes (p. 78)
16:00	Effect of outdoor access and shelter type on broiler behaviour, welfare, and performance (p. 72)	Effects of mining noise frequencies and amplitudes on behavior, physiology and welfare of wild mice (*Mus musculus*). (p. 79)
16:15	Effect of enrichment material on hair cortisol and chromogranin A in pigs (p. 73)	Behavioural responses of vertebrates to odours from other species: towards a more comprehensive model of allelochemics (p. 80)
16:45 – 18:00	AGM	
19:00	Conference dinner	

Wednesday 16 September

	Main conference hall	Conference hall
09:00-09:45	**PLENARY: Social life of captive chimpanzees: the two faces of behavioral freedom. Naruki Morimura (p. 43)**	
	SESSION 7: Freedom to express normal behaviour in captive animals	SESSION 8: Free Topics
09:50	Long-term stress monitoring of captive chimpanzees in social housings: monitoring hair cortisol level and behaviors (p. 81)	Ketoprofen reduces signs of sickness behaviour in pigs with acute respiratory disease (p. 85)
10:05	Does semi-group housing of rabbit does remove restrictions on normal behaviour? (p. 82)	Can flunixin in feed alleviate the pain associated with castration and tail docking? (p. 86)
10:20	Social housing affects the development of feeding behaviour in dairy calves (p. 83)	The effect of ketoprofen administered post farrowing on the behaviour and physiology of gilts and sows (p. 87)
10:35	Freedom to express agonistic behaviour can reduce escalated aggression between pigs (p. 84)	Is social rank associated with health of transition dairy cows? (p. 88)
10:50	Cofee Break	
11:10	Short time oral presentations (poster Group 2)	
12:00	Lunch	
13:30	Poster session Group 2 (p. 157-212)	
14:30-15:15	**PLENARY: One World-One Welfare : The Companion Animal Conundrum. Natalie Waran (p. 44)**	
	SESSION 9: Human-animal interactions and animal cognition	SESSION 10: Free Topics
15:20	Risk factors associated with stranger-directed aggression in dogs (p. 89)	Can visible eye white be used as an indicator of positive emotional state in cows? (p. 95)
15:35	Assessing stress and behaviour reactions of dogs at the veterinary clinic (p. 90)	Nasal temperatures and emotions; is there a connection in dairy cows? (p. 96)
15:50	Association of the behaviour of Swedish companion cats with husbandry and reported owner behaviours (p. 91)	The body talks: Body posture as indicator of emotional states in dairy cattle (p. 97)
16:05	Cofee Break	
16:30	Pets in the digital era: live, robot or virtual (p. 92)	Effects of intranasal administration of arginine vasopressin in Holstein steers (p. 98)
16:45	Playful handling before blood sampling improves laboratory rat affective state (p. 93)	Effect of intranasal oxytocin administration on serum oxytocin and its relationship with responses to a novel material on cow (p. 99)
17:00	The effect of visitor number on kangaroo behaviour and welfare in free-range exhibits (p. 94)	Inhaled oxytocin promotes social play in dogs (p. 100)
17:30 - 18:30	Closing ceremony	

Table of contents

Session 2. Free topics

Session 3. Free topics

Session 4. Free topics

Session 5. Freedom to express normal behaviour in captive animals

Session 6. Free topics

Session 7. Freedom to express normal behaviour in captive animals

Session 8. Free topics

Session 9. Human-animal interactions and animal cognition

Session 10. Free topics

Poster group 1

Poster group 2

Wood-Gush Memorial Lecture: Monday 14 September 8:30-10:00

A study on the normal behaviour of cattle in Japan

Shusuke Sato

Teikyo University of Science, Department of Animal Sciences, 2-2-1 Senju-Sakuragi, Adachi, Tokyo 120-0045, Japan; shusato@ntu.ac.jp

'Freedom to express normal behavior' is the most important factor for improving animal welfare. However, it is difficult to define normal behavior. 'Normal' means 'natural' in some cases, and 'instinctive' in others. In addition, Confucianism strongly influences the Japanese mindset and attaches importance to a middle way, persuades not to project, and does not advocate a primeval nature but rather a sophisticated, controlled nature. Therefore, the 'nature' respected in Japanese culture may be restricted. The Edinburgh family pen system was very influential for the Japanese mindset because it provided a sophisticated, controlled nature. First, we wrote a book, titled 'Ethograms of farm animals' in Japan, in which normal behavior was precisely categorized such that there were 80 units for cattle, 69 in pigs, 55 in chickens, and so on. Next, feral cattle on a small island in southern Japan and free-ranging cattle released into a fenceless paddock were investigated. The social structure of feral cattle was very flexible, with cattle sometimes going alone into the forest. The daily home range of free-ranging cows was 14.1 ha on average, and the average distance between daily home ranges was 462 m. These findings indicate that daily home ranges are next to each other. Third, we investigated the nature of ingestive, walking, investigative, and grooming behaviors, as well as social grouping, in order to develop an enriched production system for marbling beef. Ingestive behavior: standing grasses (to promote grazing) or harvested fresh grasses (to eliminate grazing) were available in two paddocks, for 10 minutes. We compared the speed with which the grasses were accessed by cow groups consisting of three cows each. As compared to the harvested fresh grasses (zero grazing), cows accessed the standing grasses (grazing) three and four times faster in summer and autumn, respectively. Walking: cow groups, consisting of three steers in an indoor pen (25 m^2 in size), were allowed to use an outdoor paddock (25 m^2 in size) in addition to the indoor pen for 1 h. In 1 hour, cows walked 291 m by allowing to use the outdoor paddock and 134 m only in the indoor pen. During the 22 hours after closing outdoors, cows walked an average of 4.4 km, compared to an average of 2.5 km indoors only. Investigation: In one hour, the same steers performed investigative behavior 7 and 11 times more frequently in indoor +outdoor paddocks having concrete versus soil floors, respectively, compared with investigative behaviors in an indoor pen. Grooming: Allogrooming to and manual brushing of heads and necks are solicited by receiver animals. Heart rates decreased during reception of allogrooming and manual brushing. In addition, serum oxytocin increased in the animals soon after head-brushing by people who were trusted. Social grouping: Allogrooming is exchanged between cows with familiarity and kinship. Because allogrooming is time consuming, it is impossible for cows to make bonds with many animals. We revealed that cows bond with up to approximately five animals. Based on these results, we would like to suggest a method for enriching the production system for marbling beef.

Applied ethology for ever: animal management and welfare are integral to sustainability

Michael C. Appleby

World Animal Protection, 222 Gray's Inn Road, London WC1X 8HB, United Kingdom; michaelappleby@worldanimalprotection.org

'Ethology for Sustainable Society' conveys the fact that ethology can contribute to sustainability. This review will argue, further, that applied ethology is essential for a sustainable future. Understanding the behaviour of animals is vital if they are to be managed appropriately for ecological, economic and social sustainability. These three pillars of sustainability will be considered in turn, with emphasis on 'win wins,' examples demonstrating benefits to two or more of animals, people and the environment. First, animals are a hugely important part of the environment, and behavioural studies elucidate, for example, the roles of wildlife in biodiversity and ecosystems, and animals' contributions to environmental services. Animals also impact on other aspects of the environment. Livestock make up two-thirds of terrestrial vertebrates by weight, and ethology is helping to develop approaches to management – such as sylvopastoral systems – that reduce greenhouse gas production and air and water pollution while increasing efficient use of resources and animal welfare. Second, economics concerns not just finances, but decision-making about use of resources, including resources for animals (e.g. pets) and animals as resources (e.g. livestock). Such decisions are critical worldwide for poverty eradication, livelihoods and lifestyles, and are aided by understanding animals, including their behaviour, as part of biological systems. For example, for 90% of rural households worldwide, nearly 1 billion people, animals are the main source of income. Studies of interactions between animal behaviour, health and growth – such as social behaviour affecting disease spread and vice versa – are of great benefit to both animals and people. Third, people depend on animals not only for income, food and clothing, but also for social status, security, comfort, social contact and cultural identification. Again, behaviour is central to all those roles, and understanding those relationships is fundamental to achieving the social aspects of sustainability, including reducing hunger, poverty and disease. For example, many projects on the behaviour of free-roaming dogs are improving their management, helping to eliminate rabies in both dogs and people, with obvious gains for both animal and human welfare. Sustainability cannot be achieved without addressing animal management and welfare. The most balanced decisions on animal management and welfare cannot be achieved without applied ethology. Animal management and welfare, and hence also applied ethology, are integral to a sustainable future for humanity, for the animals in our world, and for the world that sustains us all.

Animal welfare assessment: the US perspective
Cassandra Tucker
University of California, Davis, Animal Science, One Shields Ave, Davis, CA 95616, USA; cbtucker@ucdavis.edu

The largest driver of changes in US animal welfare assessment are corporate and industry-led programs. These programs aim to ensure a minimum level of care for food animals. Producer participation comes in two forms: (1) voluntary or (2) because it is required by a component of the supply chain (either a purchaser or a retailer). The most rigorous of these programs are audited by a 3rd party, someone with no connection to the farm/plant or the products produced. Outcome-based measures of animal welfare are typically included in these audits. However, well-defined facility-based measures are common in audits of species where there is a high degree of vertically integration in the business model and thus less variation in housing type or management (e.g. for laying hens, cage space/bird). In contrast, these facility-based measures are less common in audits used by sectors that are more diversified (e.g. dairy, beef). Many programs have been developed with scientific and veterinary input, and in at least some cases, have incorporated prevalence data into determination of acceptable thresholds (e.g. maximum % of a herd or flock affected by lameness). This non-regulatory approach to assessing animal welfare has a number of benefits in terms of widespread and rapid improvements and the ability to update audit criteria and thresholds on a regular basis. The challenges include: (1) the scale of assessment in terms of both number of facilities to visit and with what frequency; (2) determining key welfare criteria to include in terms of validity and feasibility; (3) ensuring consistency among auditors, and (4) the emergence of competing assurance programs within a sector. These challenges and benefits will be addressed with specific examples.

How growing environment effects on reproductive behavior – taking Orangutan (genus *Pongo*) for example

Noko Kuze

The National Museum of Nature and Science, Department of Anthropology, 4-1-1, Amakubo, Tsukuba-shi, Ibaraki, 305-0005, Japan; nouko@biglobe.jp

Reproductive success is one of important topics for 'Behavioral approaches to wild animal management'. Lack or abnormal reproductive behavior (unsuccessful mating behaver, neglect of offspring, etc.) of wildlife are usually observed under captive wildlife. This talk focus on how environment effects on reproduction of wildlife, especially mammals which have long, slow life histories. Among primates, especially great apes 'learning and experience' are very important to develop their normal behavior, included reproductive behavior. Orangutan (genus *Pongo*) is one of great apes living in tropical rainforest of South-East Asia. Orangutan is the only solitary species in extant diurnal primates and its reproductive speed is very slow; it births only one offspring every 6-9 years and age at first birth is average 15 years old in the wild. Orangutan is also important to study on wild animal management; not only a symbolic and umbrella species in conservation of tropical rainforest but also its long history – over 50 years – of 'rehabilitation/reintroduction' program in situ. In Borneo and Sumatran Island, some conservation projects have rescued orphan orangutans which lost their own mother by poaching, and reared them in captivity for several years then released in the wild since 1960s. We have studied on reproduction of female orangutan among wild, captivity and 'rehabilitant' (animals reared by human then released in the wild) then found the growing environment effects on reproductive parameters (age at first birth, inter-birth interval and infant mortality rate), reproductive behavior (mothering) and their health. For example, in rehabilitant population, average inter-birth interval is 6 years and age at first birth was average 11.6 years old, shorter and younger than wild one. Additionally in the wild orangutan, mortality rate of offspring is very low, 7-17% while in rehabilitant the mortality rate over 50%. In the wild, mothers of orangutan are hardly dead or loss of health related to pregnancy and delivery while rehabilitant mothers frequently were dead or loss of health during or after birth. In the captivity (zoos), similar high risk of health related to delivery was reported with frequent neglect. This paper reviews troubles related to pregnancy and mothering behavior in the captivity and rehabilitant orangutans compared with wild. Then I discuss what factors influence on reproduction; especially focus on growing environment of mother, rather than genetic or innate differences. Finally, I suggest what behavioral approaches are effective to improve reproduction of wild animal both in situ and ex situ management.

Social life of captive chimpanzees: the two faces of behavioral freedom

Naruki Morimura

Kyoto University, Kumamoto Sanctuary, Wildlife Research Center, 990 Ohtao, Uki City, Kumamoto, Japan, 869-3201, Japan; morimura.naruki.5a@kyoto-u.ac.jp

Primates are mammals characterized by a stunning diversity of social systems with close, enduring relationships. Some primates, especially chimpanzees (Pan troglodytes), are known to form parties with fission-fusion dynamics, defined as the variation in spatial cohesion and individual membership in a group over time. Because of large limitations on available resources in the environment, on the other hand, most captive chimpanzees have faced a significantly 'static' social life, under which they have been kept regularly in the same outdoor enclosures and night rooms, and with the same individuals of their groups. In the worst case, this continues throughout their entire lives. Thus, in the Kumamoto Sanctuary of Kyoto University, which was the first sanctuary for ex-biomedical chimpanzees in Japan, an all-male group of 15 males was formed in 2009 to stimulate social interaction among individuals as a welfare consideration. At present, the parties vary in number from 1-3 to 3-14 individuals using three adjacent outdoor compounds, in a so-called fission-fusion emulation. As a result of behavioral observation, the dynamic spatiotemporal cohesion of emulated fission-fusion successfully affected equitable and enduring bonds among those male chimpanzees. Importantly, the captive management of dynamic social life raises a further question – from the viewpoint of behavioral freedom – that remains unanswered for any agency that makes a choice to form a party with fission-fusion dynamics. It has been pointed out that wild animals in captivity, including chimpanzees, are unique species in welfare consideration, which allows us to compare their wild and captive conditions directly. Given that wild chimpanzees are facing high ecological pressures in fragmented habitat as they compete for land with humans, the degree of behavioral freedom is also an issue that is shared among chimpanzees in both conditions. Understanding how wild chimpanzees solve a problem derived from the limitations in their environment, or in the individual, that prevent self-initiated choice and action may give us new insight into behavioral freedom. The field experiment of tool-making behavior in chimpanzees at Bossou, Guinea, revealed that a 57-year-old female chimpanzee with left-arm paralysis made a leaf tool for drinking water from a tree hollow by the hierarchical combination of manual-pedal motor actions including the left arm. Therefore, the comparative cognitive study of wild animals in both captivity and the wild on overcoming their physical, ecological, and social limitations can possibly provide us an opportunity to study behavioral flexibility that reflects cognitive competences as the evolutionary bases of behavioral freedom that is comparable to humans.

One world-one welfare: the companion animal conundrum

Natalie Waran
University of Edinburgh, Jeanne Marchig International Animal Welfare Centre, RDSVS, Easter Bush, Midlothian, Scotland UK, EH25 9RG, United Kingdom; natalie.waran@ed.ac.uk

Humanity is faced with a range of global animal related challenges to which scientific knowledge should be applied. Providing food security universally is one such challenge and alongside the need for greater animal production, is a growing awareness and understanding of the critical relationship between poor standards of animal health (and productivity) with reduced animal welfare, and the important link with the health of humans. Animals and humans share the same environments, the same pathogens and similar disease processes. Improving the lives of humans and animals, relies on understanding the interconnectedness of species, a greater focus on preventative measures, and the importance of translating relevant research into practical solutions. There is increasing recognition that emerging infectious diseases are associated with increasing contact and conflict between people and animals. This can be seen when we consider the increasing numbers of companion animals living in close proximity with humans, and concerns regarding the spread of zoonotic diseases such as toxoplasmosis and rabies. In addition to such health issues, are rising concerns for the physical and psychological health of individual animals, and the challenge of minimising the suffering of animals over which we have a duty of care. In this presentation I will explore the conundrums, conflicts and challenges between humans and companion animals whether these be kept as pets or those living as stray/street animals. I will consider the 'real world' welfare issues experienced by 'companion animals' in different parts of the world, and why the development of evidence based, robust approaches for assessing their welfare are needed, to enable us to balance the need for solutions to public health issues such as an increasing stray/street dog population, and dog bites in the home, with a need to protect the welfare of individual animals.

Producer estimations and perceived importance of lameness in dairy cattle: where do we move from here?

Janet H Higginson Cutler[1], Jeff Rushen[2], Anne Marie De Passillé[2], Jenny Gibbons[3], Karin Orsel[4], Ed A Pajor[4], Herman W Barkema[4], Laura Solano[4], Doris Pellerin[5], Derek Haley[1] and Elsa Vasseur[6]

[1]University of Guelph, Ontario Veterinary College, Guelph, Ontario, N1G 2W1, Canada, [2]University of British Columbia, Dairy Research and Education Center, Agassiz, British Columbia, V0M 1A0, Canada, [3]DairyCo, Agriculture & Horticulture Development Board, Stoneleigh Park, Kenilworth, Warks, CV8 2TL, United Kingdom, [4]University of Calgary, Department of Production Animal Health, Calgary, Alberta, T2N 4N1, Canada, [5]Université Laval, Département des sciences animales, Québec, Québec, G1V 0A6, Canada, [6]University of Guelph, Organic Dairy Research Center, Alfred, Ontario, K0B 1A0, Canada; vasseur.elsa@gmail.com

Lameness is one of the most important welfare and productivity concerns in the dairy industry. The objectives of our study were to obtain producer prevalence estimates and perceptions of lameness, and to investigate the method used by producers to monitor lameness in their herds. During a farm visit on 237 Canadian dairy herds, on each farm 40 focal cows were scored for lameness (obviously lame or sound) by trained researchers either while walking (obvious limp or sound) or standing in the tie-stall (2 SLS indicators or sound). In order to provide some explanation of challenges in lameness control in the surveyed herds, producers were interviewed on the same day using a survey questionnaire. Median farm-level prevalence estimated by producers was 5.5% (range 0-50), whereas the researchers observed a median prevalence of 20.0% (range 2.5-69). Correlation between producer and researcher estimated lameness prevalence was low (r spearman=0.19; P=0.005). When asked about lameness severity, 48% of producers thought lameness was a moderate or major problem in their herds. One third of producers considered lameness as the highest ranked health problem they were trying to control while two thirds of producers stated that they made management changes to deal with lameness in the past 2 yrs. Almost all producers (98%) stated they routinely checked cows to identify new cases of lameness; however, 40% of producers did not keep records of lameness. The underestimation of lameness prevalence by producers highlights that there continues to be difficulty in lameness detection on Canadian dairy herds. Producers are aware of lameness as an issue in their herds and almost all are monitoring for lameness as part of their daily routine. Training to improve detection methods is likely to help decrease lameness prevalence.

The influence of gentle interactions during milking on avoidance distance, milk quality and milk flow in dairy cows

Stephanie Lürzel[1], Kerstin Barth[2], Andreas Futschik[3] and Susanne Waiblinger[1]
[1]*Institute of Animal Husbandry and Animal Welfare, University of Veterinary Medicine, Vienna, Veterinärplatz 1, 1210 Vienna, Austria,* [2]*Johann Heinrich von Thünen-Institute, Institute of Organic Farming, Trenthorst 32, 23847 Westerau, Germany,* [3]*Department of Applied Statistics, JK University Linz, Altenberger Str. 69, 4040 Linz, Austria; stephanie.luerzel@vetmeduni.ac.at*

We investigated a potential connection between the animal-human relationship and milk quality parameters in 26 German Holstein cows. 14 cows were handled gently (stroking, talking) for 2 min during both milkings on 15 d, totalling 60 min per cow. The same experimenter remained standing in a similar distance to the control cows for the same amount of time. Before and after the treatment phase, the experimenter measured avoidance distances at the feeding rack, and we took milk samples in order to analyse milk quality parameters (fat, protein, lactose, fat-free dry matter, somatic cell count). We recorded milk flow parameters and milk yield before, during and after the treatment phase. Avoidance distances decreased in the stroked cows (before: median 15 cm (range 0-70 cm); after: 3 (0-45) cm, Wilcoxon: $P=0.002$) but not in the controls (before: 8 (0-60) cm; after: 8 (0-26) cm; $P=0.45$). There was no effect of treatment on milk yield per milking (stroked: mean 11.9 kg, control: 13.0 kg, GLMM: $P=0.45$) or peak flow rate (stroked: 3.3 kg/min, control: 3.2 kg/min, GLMM: $P=0.71$). Bimodalities occurred more often in the stroked cows than in the controls on the first day of stroking (stroked: 8/14, control: 1/12, Fisher's exact test: $P=0.01$). There was no effect of treatment on any of the milk quality parameters. The stroking treatment was effective in decreasing the avoidance distances, which indicates an improved animal-human relationship. However, this improvement did not have any effects on milk quality, milk yield or milk flow parameters. The relationship between milkers and cows may have obscured possible effects. The impairment of milk ejection on the first day of the stroking treatment may be due to the novelty of the situation.

Human-animal interactions and safety during dairy cattle handling

Cecilia Lindahl[1], Stefan Pinzke[2], Anders Herlin[3] and Linda J Keeling[4]
[1]Swedish Institute of Agricultural and Environmental Engineering, P.O. Box 7033, 750 07 Uppsala, Sweden, [2]Swedish University of Agricultural Sciences, Department of Work Science, Business Economics and Environmental Psychology, P.O. Box 88, 230 53 Alnarp, Sweden, [3]Swedish University of Agricultural Sciences, Department of Biosystems and Technology, P.O. Box 103, 230 53 Alnarp, Sweden, [4]Swedish University of Agricultural Sciences, Department of Animal Environment and Health, P.O. Box 7068, 750 07 Uppsala, Sweden; cecilia.lindahl@jti.se

When cattle perceive situations as aversive, they may become agitated and potentially dangerous to handle. This study aimed to investigate the effect of frequency of a routine on cow-handler interactions and handler safety by comparing moving cows to daily milking and moving cows to more rarely occurring hoof trimming. The study was conducted on 12 Swedish commercial dairy farms. Each farm was visited twice, once to observe cows being collected and moved to milking and once to hoof trimming, with the same handler on each farm being observed on the two visits. The study included behavioural observations of handlers and cows, heart rate measurements (HR) of cows and recording of incidents (defined as physical contact between cow and handler that could have resulted in an injury, i.e. kicked, head butted). At milking, cows were easily moved with few interactions by the handler. As expected, the cows showed no behavioural signs of stress, fear or resistance and their HR only rose slightly from baseline (their average HR during an undisturbed period before handling). When moving cows to hoof trimming, higher proportions of gentle and forceful interactions where used than when moving them to milking (Wilcoxon signed rank test; $P<0.01$). Furthermore, the cows showed significantly higher frequencies of behaviours indicative of aversion and fear, e.g. freezing, balking, and resistance, as well as a higher increase in HR (paired t-test; $P<0.001$). No incidents were observed when cows were moved to milking whereas during moving to hoof trimming, the frequency of incidents ranged from none to 0.1 per min (median 0.03 per min). Some interactions were found to be correlated to incidents, i.e. forceful tactile interactions with an object were correlated to the handler being kicked ($rs=0.76$, $P<0.01$). The results indicated a very high injury risk to the handler during the hoof trimming procedure which usually occurs only 2-3 times a year. Furthermore, during hoof trimming cows showed more behaviours indicative of aversion and fear and were more difficult to move causing the handlers to use force. These results indicate a need for changes in the way procedures that may be perceived as aversive to the cows are performed on dairy farms to increase handler safety and improve cow welfare, ease of handling and efficiency.

Daily lying behaviour as an indicator of pregnancy toxemia in dairy goats

G. Zobel[1], K. Leslie[2], D.M. Weary[1] and M.A.G. Von Keyserlingk[1]
[1]*University of British Columbia, Animal Welfare Program, 180-2357 Main Mall, Vancouver, BC, V6E1V4, Canada,* [2]*University of Guelph, Population Medicine, 50 Stone Road, Guelph, ON, N1G 2W1, Canada; g_zobel@yahoo.ca*

When clinical signs such as lethargy and immobility are noted, the prognosis for recovery from pregnancy toxemia is poor. These behavioural changes likely occur progressively, and could therefore be used as early indicators of illness. The aim was to evaluate the value of monitoring lying behaviour to identify illness. Daily lying time and lying bout frequency were calculated for 12 d before kidding on 10 commercial dairy goat farms in Ontario, Canada. Does were monitored for ketonemia (elevated blood β-Hydroxybutyrate, BHBA), a factor associated with pregnancy toxemia. Does were considered healthy (n=232) when BHBA <0.9 mmol/l before and after kidding, and ketonemic (n=14) when BHBA ≥1.7 mmol/l before kidding. PROC GLIMMIX (SAS) models were used to assess the effect of health status on lying time and lying bout frequency, with litter size as a covariate. Results presented as mean ± SED (lying time) and mean and 95% CI (lying bout frequency). Ketonemic does lay longer than healthy does (15.5 vs 12.8 h/d, SED=0.9; P=0.002). Lying time decreased near kidding, but compared to healthy does, ketonemic does continued to lie down longer on the day before kidding (16.0 vs 13.2 h/d, SED=1.1; P=0.02) and on kidding day (13.0 vs 9.8 h/d, SED=1.1; P=0.005). Does carrying triplets tended to have longer lying times by 1.0 h/d (SED=0.3; P=0.07). There was no difference in lying bout frequency between ketonemic and healthy does; all does increased the number of lying bouts between the day before kidding and kidding day (16.8 (15.8-17.8) vs 20.5 (19.4-21.8) bout/d; P<0.0001)). In summary, ketonemic goats continued to lie down around kidding time, a period normally associated with increased restlessness. Behavioural differences were noted throughout the 12 d monitoring period suggesting that at-risk goats could be identified well in advance of kidding using lying time monitoring.

What can carcass-based assessments tell us about the lifetime welfare of pigs?

Grace Carroll[1], Laura Boyle[2], Alison Hanlon[3], Kym Griffin[1], Lisa Collins[4] and Niamh O' Connell[1]
[1]Queens University Belfast, Biological Sciences, University Road, Belfast, BT7 1NN, Northern Ireland, United Kingdom, [2]Teagasc, Animal & Grassland Research & Innovation Centre, Moorepark, Fermoy, Co. Cork, Ireland, [3]University College Dublin, School of Veterinary Medicine, Belfield, Dublin 4, Ireland, [4]University of Lincoln, School of Life Sciences, Brayford Pool, Lincoln, Lincolnshire, LN6 7TS, England, United Kingdom; gcarroll05@qub.ac.uk

There is increasing interest in developing abattoir-based measures of farm animal welfare. However, it is important to understand the extent to which these measures reflect lifetime welfare. The study aim was to determine the extent to which on-farm health and welfare issues were reflected in measures taken from the carcass. Data were collected from May 2013 to December 2014 from ten batches of pigs (n=720) reared in conventional intensive housing. All pigs had 50% of the tail length docked within 24 hours of birth. Each animal was assessed at 7, 9 and 10 weeks of age (early life), and at 15 and 20 weeks of age (later life). At each time-point, pigs were assessed for tail lesions (TL [0 to 5 scale considering the extent of damage and loss of tail length]), skin lesions (SL [0 to 5 scale considering the number and severity of lesions]), and health issues (HI) (namely; lameness, bursitis, coughing, scouring, rectal prolapse, hernias and aural hematomas). Following slaughter at 21 weeks, each carcass was scored for tail length (long ≥6 cm, short ≤5 cm), TL and fresh (red) and old (non-red) SL. Using a Kruskal-Wallis ANOVA, the carcass measures of pigs with TL, SL and HI in 'early life' (EL), 'later life' (LL) or 'whole life' (WL) were compared to their respective controls (C) which did not have TL, SL or HI at any time point on farm. SL pigs had more healed carcass SL than C pigs when SL were recorded in EL, LL ($P<0.05$) or WL ($P<0.001$). Fresh SL to the carcass did not differ between groups. Pigs recorded as having TL in EL ($P<0.05$), LL or WL ($P<0.001$) had more carcass TL than C pigs. Pigs recorded as having TL in LL and WL ($P<0.001$), but not during EL ($P>0.05$), had shorter tails on the carcass than C pigs. HI were not reflected in carcass measures ($P>0.05$). These findings show that carcass-based measures of skin and tail lesions can be used to detect welfare problems occurring during both the growing and finishing period of pigs' lives. Fresh skin lesions may reflect damage that occured during the marketing process.

From farm to slaughter: broiler welfare assessment during catching, loading, transport and lairage

Leonie Jacobs[1,2], Evelyne Delezie[2], Luc Duchateau[1], Xavier Gellynck[1], Klara Goethals[1], Evelien Lambrecht[1] and Frank Tuyttens[1,2]
[1]*Ghent University, Salisburylaan 133, 9820 Merelbeke, Belgium,* [2]*Institute for Agricultural and Fisheries Research (ILVO), Scheldeweg 68, 9090 Melle, Belgium; leonie.jacobs@ilvo.vlaanderen.be*

The transport process (starting with catching on farm and ending with unloading and lairage at the slaughter plant) is a critical and stressful phase of the production process, with potentially serious animal welfare and economic implications. This study's aim was to assess the impact of various phases of the transport process (catching/loading, transport and lairage), and their related risk factors, on broiler welfare. A protocol with 19 animal-based measures was used to assess broiler welfare during transport, with measures based on EFSA recommendations and Welfare Quality®. Welfare was assessed for 82 transports from 52 commercial farms (flock size range 13,750-160,000) to five slaughter plants in Belgium. Each transport was assessed at four moments: on farm before catching/loading, on farm after catching/loading, at the slaughter plant during lairage, and after slaughter. Data were log-transformed (except for body weight and temperature) and analysed in SAS using PROC MIXED and GLIMMIX with farm as a random effect. Results are shown as mean±SE. Broilers lost 5.3% of their weight during transport (2.65±0.01 kg before catching, 2.51±0.01 kg after transport; $P<0.001$). Weight loss tended to increase with transport duration (43±22 g loss/h, range 0.5-5 h; $P=0.074$). Body temperature was lower after transport compared to after catching/loading (40.5±0.01 vs 41.2±0.01 °C; $P<0.001$). The prevalence of birds with wing fractures was higher after catching and loading than before catching (1.75 vs 0.08%; $P<0.01$). Dead on Arrivals (DOAs) were affected by type of catch crew: fewer DOAs were found when the birds had been caught and loaded by a professional team compared to a team consisting of friends or family (0.23±0.03% vs 0.46±0.12%; $P=0.001$). There were fewer carcass rejections at slaughter when transports occurred at night (0.49±0.05%) compared to morning (1.07±0.24%; $P=0.025$) or daytime (1.42±0.33%; $P<0.001$). Each increase of ambient temperature with 1 °C (range 0.1-25.7 °C) resulted in a 1.21-fold increase of broilers with bruised legs ($P=0.037$) and a 1.02-fold increase of broilers with bruised breasts and wings ($P=0.027$). No significant associations between the lairage phase and welfare parameters were found in this study. To conclude, a number of risk factors (transport duration, type of catchers, moment of transport, ambient temperature) were identified for weight loss, DOAs, carcass rejections and bruises. Also, critical transport phases (catching/loading and transport) for body temperature and fractures have been identified.

Relationship between lameness scored on farm and foot pad lesions scored at the abattoir in broiler chicken flocks

Kathe Elise Kittelsen[1], Randi Oppermann Moe[2], Erik Georg Granquist[2], Jens Askov Jensen[3] and Bruce David[4]
[1]Animalia, Norwegian Meat and Poultry Research Centre, Lørenveien 38, 0515 Oslo, Norway, [2]Norwegian University of Life Sciences, Faculty of Veterinary Medicine and Biosciences, Ullevålsvn 66, 0033 Oslo, Norway, [3]University of Aarhus, Blichers Allé 20, 8830 Tjele, Denmark, [4]Norwegian Veterinary Institute, Department of pathology, Ullevålsveien 68, 0454 Oslo, Norway; kathe.kittelsen@animalia.no

Lameness and the occurrence of foot pad lesions in rapidly growing birds are welfare and economical concerns in broiler chicken production worldwide. In general, the herd prevalence of leg pathologies increases towards the end of the growth period. In Norway, scoring of foot pad lesions at the abattoir is used as a tool to assess animal welfare in all broiler flocks. Scores achieved have consequences for maximum allowance of animal densities for future flocks on the farm concerned. The aim of this study was to investigate the general leg health in Norwegian broiler flocks close to slaughter age by using gait scores and foot pad lesion scores as predictors, and to identify the strength of relationship between these two measures of leg health. A total of 60 broiler flocks were included and 100 birds were scored in each flock (i.e. 10% of all Norwegian broiler farms). For evaluation of lameness, birds were inspected visually to obtain a gait score ranging from 0 to 5, where score 0 was no discernible limp and 5 was a complete inability to walk. A gait-score (GS) of 2 and above is a definable gait defect. GS of 3 and above interferes with the birds' ability to maneuver. Scoring of individual birds took between 10 seconds and 1 minute. At the abattoirs, total foot pad lesion scores for each flock were registered in 100 birds on a scale ranging from 0-200. A score between 0-80 are defined as satisfactory, 80-120 are non-satisfactory and scores above 120 are unacceptable. The mean footpad-lesion score was 14.86 (range: 0-100) and mean gait score was 1.69 (range 1.02-3.21). Correlation analyses showed a weak but not significant relationship between gait scores and foot pad lesion scores ($r=0.17$, $P=0.27$). The bivariate regression showed that foot pad lesion scores for the sample population only explained 2.7% of the mean gait scores ($r2=0.027$, $P=0.27$). However, 17% of the broilers had gait scores that represent painful conditions (GS>2). The findings indicate that foot pad lesion scores alone do not reflect the total welfare situation in regards to broiler leg health on a flock level. In conclusion, lameness scoring on farm should be included as an additional measure in future broiler welfare assessment systems.

Reliability of the visual gait score as a method of gait assessment in domestic ducks
Brendan Duggan, Paul Hocking and Dylan Clements
The Roslin Institute, University of Edinburgh, Easter Bush Campus, Midlothian, EH25 9RG, United Kingdom; brendan.duggan@roslin.ed.ac.uk

Intensively reared poultry often suffer from gait abnormalities. The aim of this study was to evaluate the reliability of the current visual gait scoring technique as a method of assessing gait in Pekin ducks and to compare this with objective gait measurements gathered using a pressure platform. Birds were assessed by four industry gait scorers at three, five and seven weeks of age. Scorers were shown three video sequences of 36 birds at each age walking over a runway. Each of the three video sequences contained 144 walks – 4 walks (including one duplicate) from each bird. Scorers were asked to rate each walk with a score of 1 (poor) to 5 (perfect gait). None of the scorers were informed that the sequences contained duplicate recordings. At 7 weeks, when scoring the same walks, all 4 scorers agreed 28% of the time and 3 of the 4 scorers agreed 74% of the time. Individual scorers failed to allocate the same score to two duplicate walks 26% of the time. A mixed model using restricted maximum likelihood showed that, on average, observers varied between each other by 0.39 of a gait score and individual observers varied by 0.55 when scoring two duplicate walks. No clear observer drift effect was detected – scorers deviated to a similar degree when scoring the first 60 walks compared to the last 60. The results demonstrate that this gait scoring system is not a reliable method of assessing gait in ducks in this scenario. A consensus score of all scorers will be used to compare the visual scoring system to an objective system using a pressure platform. The end goal of this study is to develop a more objective method of assessing gait in poultry which will inform industry breeding decisions.

Startling pigs – a new measure of pig affect and welfare

Poppy Statham[1], Neill Campbell[2], Sion Hannuna[2], Samantha Jones[1], Bethany Loftus[1], Jo Murrell[1], Emma Robinson[3], Elizabeth Paul[1], William Browne[4] and Michael Mendl[1]
[1]University of Bristol, School of Veterinary Science, Langford House, Langford, Bristol, BS40 5DU, United Kingdom, [2]University of Bristol, Department of Computing Science, Merchant Venturers Building, Woodland Rd, Bristol, BS8 1UB, United Kingdom, [3]University of Bristol, Physiology and Pharmacology, Medical Sciences Building, University Walk, Bristol, BS8 1TD, United Kingdom, [4]University of Bristol, Centre for Multilevel Modelling, Graduate School for Education, 2 Priory Road, Bristol, BS8 1TX, United Kingdom; poppy.statham@bristol.ac.uk

A truly accurate assessment of on-farm welfare requires validated proxy measures of animal affective states. In humans and rodents, the 'Defence Cascade' (DC) response to startling stimuli appears to reflect affective state. We investigated whether the same applies to pigs and whether the DC response is therefore a potential on-farm welfare measure. Twelve groups of four pigs were placed in either 'Barren' or 'Enriched' housing. After six weeks, the treatments were switched. In week 3 in each environment 'Initial' DC sessions were completed (no drug administration or saliva collection); the pigs were filmed in their home pen whilst a startling stimulus (varied due to habituation) was applied three times. In weeks 4 to 6, oral doses of Diazepam (0.3 mg/kg), Reboxetine (1 mg/kg) or Control (unadulterated food) were administered prior to DC testing, balancing order across groups. A trained observer rated startle magnitudes (SM) on a scale from 0 to 4, and recorded any freeze responses. Saliva samples for cortisol analysis (SC) were collected before drug administration and after testing. Multilevel models were used to account for the data structure; ordered multinomial for SM, binomial for freeze, and normal for SC. Wald test statistics are presented. 'Barren' pigs showed decreased SM (χ^2=21.923, P<0.0001) probability of freezing (χ^2=21.158, P<0.0001) and SC (χ^2=6.524, P<0.05) compared to 'Enriched' pigs. Pigs treated with Diazepam showed decreased SM (χ^2=6.352, P<0.05) compared to Control, and a decreased probability of freezing (χ^2=15.797, P<0.0001) and lower SC (χ^2=27.657, P<0.0001) compared to Reboxetine. The DC response was reduced with Diazepam (anxiolytic) and increased with an acute dose of Reboxetine (anxiogenic in humans). Environmental manipulations did not change DC responses as expected, and possible explanations for this will be presented. DC responses matched SC results. Overall, the DC response shows promise as a new indicator of affect and welfare in pigs.

Repeated testing in a cognitive bias paradigm in pigs

Jenny Stracke, Sandra Düpjan, Armin Tuchscherer and Birger Puppe
Leibniz Institute of Farm Animal Biology, Behavioural Physiology, Wilhelm-Stahl-Allee 2, 18196
Dummerstorf, Germany; stracke@fbn-dummerstorf.de

The cognitive bias approach provides information on the valence (positive/negative) of affective states in non-human animals. It is based on a learning process where animals learn to distinguish two different stimuli. This information has to be transferred to interpret the impact of novel, ambivalent stimuli. Learning and neural consolidation play a crucial role in enabling repeated testing which, up to now, was impossible in pigs. We here present a design which enables up to eight test repetitions, a prerequisite to testing before/after treatments and analysing stable individual response patterns. Subjects (26) underwent a 7-day-training, learning to associate the spatial position of a goal-box either positively or negatively (operant conditioning with partial reinforcement) in a go/no-go paradigm. Training was followed by four weeks of testing where one group of animals (n=15) had a one-day break between training and tests (group A), whereas the other group (n=11) were allowed three days of consolidation (group B). In the four test-weeks, the goal box was additionally presented on three unreinforced, ambivalent test positions, each tested eight times. Latency to open the goal box was measured and an individual relative latency was calculated for each animal/day. Data were analysed using mixed effect models including repeated measurements and post-hoc Tukey Kramer t-tests. Subjects learned to discriminate S+ and S- reliably ($P<0.001$; F=5,815.3). We could show a graded response to the ambivalent test positions ($P<0.001$; F=509.4). Whereas group A showed significant differences in latency between test week 1 and the remaining three weeks ($P<0.01$; F=4.5), group B displayed stable latencies over the whole 4 weeks. In conclusion, the modified paradigm for repeated testing for spatial judgement biases in domestic pigs will enable future studies not only on treatment induced affective states in animals but on optimism/pessimism as personality traits.

Is the cognitive bias test able to assess changes in emotional state due to housing conditions in pigs?

Ricard Carreras[1], Eva Mainau[1,2], Xènia Moles[1], Antoni Dalmau[1], Xavier Manteca[2] and Antonio Velarde[1]
[1]IRTA, Animal Welfare Subprogram, Veïnat de Sies, s/n, 17121 Monells, Spain, [2]Universitat Autònoma de Barcelona, Department of Animal and Food Science, School of Veterinary Science, 08193 Bellaterra, Spain; ricard.carreras@irta.cat

The aim of the study was to assess the effect of the housing conditions on cognitive bias (CB) in pigs. 44 female pigs of 8 weeks of age were housed in four pens. During the first 7 weeks, pigs were allocated under the same housing conditions (fully slatted floor with a density of 1.2 m^2/pig). The following 7 weeks, the density of two pens was increased to 0.7 m^2/pig (barren) whereas in the other two pens the floor was changed to concrete and 700 g of straw/pig were provided every 2-3 days (enriched). Three CB trials were performed: the first one before the change of the housing conditions, and the second and third were done after 1 and 5 weeks, respectively. Before the trials pigs were trained during 11 days to discriminate between a bucket with apples (reward; R) and a bucket with air puff (punishment; P) according to its position (left or right) in the test pen. Afterwards, each animal was subjected to the trials consisting in three types of test, one in the middle between R and P (M test), one between the R and the middle (R test) and the other between the P and the middle (P test). The time to cross a line marked on the floor 1 m around the bucket was recorded. No significant effect was found between groups during the three trials in the different tests, with the exception of the R test during the second (P=0.001) and third trial (P=0.024). However during the second trial the enriched group spent more time to cross the line, while during the third trial the barren group took more time than the others. The lack of effect of housing conditions on CB could be due to the low intensity of the treatments, the low sensitivity of the technique or both factors together.

The association between plumage condition, fearfulness and range use as monitored by RFID, in free-range laying hens

Kate Hartcher[1], Kelly Hickey[1], Paul Hemsworth[2], Greg Cronin[1] and Mini Singh[1]
[1]*University of Sydney, Poultry Research Foundation, Veterinary Faculty, 425 Werombi Road Camden, NSW 2570, Australia,* [2]*University of Melbourne, Animal Welfare Science Centre, Faculty of Veterinary and Agricultural Sciences, Parkville, VIC 3010, Australia; kate.hartcher@sydney.edu.au*

Severe feather-pecking, a particularly detrimental behaviour in laying hens, is thought to be negatively correlated with range use in free-range systems. In turn, range use is thought to be associated with fearfulness, where fearful birds are less likely to venture outside. This experiment aimed to investigate the associations between range use, fearfulness and plumage condition. Two pens of 50 ISA Brown laying hens, housed in an experimental facility, were fitted with RFID transponders (contained within rubber leg rings) at 26 weeks of age. Individual range use was then recorded for 13 days, and all birds were individually weighed and feather-scored at the beginning and end of the trial. Thirty birds were selected for tonic immobility testing. Ninety-six percent of all birds used the outdoor run, with an average total duration of 80.95 h on range per bird, and an average of 192 visits. There was no association between feather-scores and time spent outside ($P=0.76$) or visits to the range ($P=0.93$). There was also no association between tonic immobility duration and time spent outside ($P=0.47$), and a trend for number of visits ($P=0.06$). A possible explanation for the lack of relationships is that there was little variation between birds. That is, fearful birds may be less likely to use the range, but the birds in the present experiment had similar levels of fearfulness, as assessed by tonic immobility durations ($P=0.32$). This could be due to the small group sizes and regular handling. Plumage damage was also similar between birds ($P=0.18$). The RFID technology was effective in collecting large amounts of data on range use in the tagged birds, and provides a means for quantitatively assessing range use in laying hens. The Australian Egg Corporation Limited is acknowledged for providing funding for the experimental setup.

Auto-detecting cattle behaviour using IR depth-sensor camera

Daisuke Kohari[1], Tsuyoshi Okayama[1], Haruka Sato[1], Kasumi Matsuo[1], Takami Kosako[2], Tatsuhiko Goto[1] and Atsushi Toyoda[1]
[1]Ibaraki University, College of Agriculture, 4668-1, ami-machi, inashiki-gun, Ibaraki, 3000331, Japan, [2]NARO Institute of Livestock and Grassland Science (NILGS), 2 Ikenodai, Tsukuba, Ibaraki, 3050901, Japan; kohari@mx.ibaraki.ac.jp

With the recent development of hardware such as various cameras and sensors, methods of observing animal behaviour are becoming increasingly precise. An IR depth-sensor camera developed for use with video games can produce information about various 3-D behaviour of subjects according to depth changes in the captured area. In this study, we tried auto-detecting cattle behaviour using an IR depth-sensor camera. Observations were conducted in a parturition pen (w3×d4 m) at the Field Science Center of Ibaraki University. Three breeding cattle subject animals were examined immediately before parturition. An IR depth-sensor camera (Xtion PROLIVE; ASUSTek Computer Inc.) placed 4 m above the center of the experimental pen recorded the whole parturition pen area. Although this sensor camera can record depth data (320×240 pixel) at 60 fps, the cattle behaviour was recorded at the rate of 1 frame each 5 s because of the large volume of data. Recorded data were stored on external HDD via a laptop computer. We observed two 24 h periods of pre-parturition behaviour of cattle such as posture change (standing, lying down) and tail lifting. Each behaviour was transformed from the recorded depth data to graphical images using software (Processing ver. 2.2.1; Processing. org), and was compared with related depth information or the difference of each graphical documents. Differences between standing and lying down of cattle were discerned easily from depth information (standing, 1,391±44 mm; lying, 812±75 mm; t-test $P<0.01$). On the other hand, tail lift was assessed using the difference of their longitudinal length of graphical image during lifting/ not-lifting the tail (lifted, 6.6±4.3 pixel; not lifted, 41.1±3.0 pixel; t-test $P<0.01$). In conclusion, automatic detection of cattle behaviour was accomplished using an IR depth-sensor camera. Detection algorithms appropriate for the characteristics of respective behaviours should be designed.

Evaluating health of jungle crow using animal-borne accelerometers

Tsutomu Takeda[1], Katsufumi Sato[2] and Shoei Sugita[1]
[1]Utsunomiya University, Department of Agriculture, 350 Mine, Utsunomiya, Tochigi, 321-8505, Japan, [2]Atmosphere and Ocean Research Institute, The University of Tokyo, Behavior, Ecology and Observation Systems, 5-1-5 Kashiwanoha, Kashiwa, Chiba, 277-8564, Japan; ixodes@cc.utsunomiya-u.ac.jp

The prevention of infectious diseases, such as avian influenza, in poultry farms or zoos is of primary importance in the care and management of these animals. The detection of the initial symptoms can give time to plan more efficient control and prevention measures to avoid further spread of the disease. In this study, we observed behaviors of jungle crows and correlated them with the blood hematocrit (Ht) values. Six wild crows were kept in two cages 2.0×2.5×2.5 m in April and 0.1 ml blood was collected. Ht values were measured weekly until the end of October 2014. Additionally, three-axis accelerometers (OR1400-D3GT) were deployed on the back of crows using a 60×30 mm cotton harness with the hook and loop fastener. The total mass of the harness with an accelerometer was 16 g which was less than 2.5% of crow body mass. Handling time was less than 5 minutes in each bird. Accelerometers recorded three-dimensional acceleration (0.05 seconds interval, respectively) for 50 hours. The measurements were conducted twice per week for six months. Dynamic movements of crow was evaluated by Overall Dynamic Body Acceleration (ODBA), the sum of the absolute values of the dynamic accelerations from all three axes. And longitudinal accelerations of all birds were used to calculate Power Spectral Densities (PSD) using Fast Fourier Transformation to characterize the behavior, which may be related to physucal conditions. All procedures were performed in accordance with the ethical guidelines (No. A10-007). Ht values of all crows were around 50% through first two months. Values of five crows were kept around similar values until the end of experiment. However, Ht value of one crow started decreasing from July 11 and the crow died on October 10. The last Ht value was 8% on October 9. Body mass and rectal temperatures of this crow also continually decreased. Weekly average values of Ht had a positive correlation (r=0.77) with ODBA, which indicate the crow was active during daytime when its Ht was high. Furthermore, in this crow's PSD the peak at 5.6 Hz appeared more frequent in comparison with PSDs of other five healthy crows. The rate of the peak appearance was 41% in the dead crow, however, those of the others were 15±6% in average. According to video observation of five crows, the periodical movement of 5.6 Hz in swaying acceleration seemed to correspond to shivering movements of their bodies. It was suggested that jungle crows shivered to remove ectoparasites as they became worse physical condition including the anemia.

Relations between play, vocalization and energy intake of dairy calves following separation from the cow

Jeffrey Rushen[1], Rebecca Wright[1], Julie Johnson[2], Cecilie Mejdell[2] and Anne Marie De Passille[1]
[1]University of British Columbia, Faculty of Land and Food Systems, Highway 7, Agassiz, BC V0M 1A0 BC, Canada, [2]Norwegian Veterinary Institute, Dep. Of Health Surveillance, P.O. Box 750, Oslo, Norway; jeffrushen@gmail.com

Play behaviour is an indicator of good welfare in young calves and is reduced by low energy intake and weaning off milk. There is renewed interest in keeping calves longer with the cow but separation leads to signs of distress, such as vocalizations. Providing calves with an alternative milk source prior to separation may help them adapt to the separation. We hypothesized that (1) locomotor play of calves reared with their mothers will reflect their energy intake after separation, and (2) calves' prior access to automated milk feeders will increase energy intake and play and reduce vocalizations after separation. 30 Holstein calves and their dams were kept in adjacent pens. 'Suckling-only' calves were allowed to suckle the cow during the night and received no other milk while 'Suckling-and-milk-feeder' calves could suckle during the night and were allowed 12 l/d of milk from an automated milk feeder. At 6 weeks of age, calves were not allowed to enter the cow pen during the night but had access to automated milk and grain feeders. To measure locomotor play, we placed the calves individually in a 9.5×2 m arena for 10 min, during 4 d before and 3 d after separation. The frequency of jumping and vocalization were scored. Digestible energy (DE) intakes were calculated from combined milk and grain intake. Before separation, there were no differences between the two treatment groups on any behavioural measure ($P>0.10$). After separation, suckle-only calves had a lower frequency of jumping (Mann-Whitney test, $P=0.02$) and tended to vocalize more frequently (Mann-Whitney test, $P=0.08$) than did suckle-and-milk-feeder calves. For all calves, the number of vocalization was negatively correlated with frequency of jumping ($r=-0.51$; $P=0.005$). After separation, suckle-only calves had lower energy intakes than suckle-and-milk-feeder calves (Mann-Whitney test, $P=0.01$) and the digestible energy intake of the calves was positively correlated with the frequency of jumping ($r=0.75$; $P<0.001$), and negatively correlated with the frequency of vocalization ($r=-0.59$; $P<0.001$). Our results show that a low energy intake after separation from the dams is associated with reduced locomotor play and increased vocalization and that prior access to an automated milk feeder helps maintain energy intake after separation. The association between vocalization and locomotor play indicates that the reduction in play at separation is related to the emotional response of the calves.

Weaning calves off milk according to their ability to eat solid feed reduces hunger and improves weight gain during weaning

Anne Marie De Passille and Jeffrey Rushen
University of British Columbia, Faculty of Land and Food Systems, Highway 7, Agassiz, BC, V0M 1A0, Canada; amdepassille@gmail.com

Dairy calves weaned off milk show signs of hunger and can lose weight. We examined if weaning calves according to solid feed intake reduces the effects of weaning. Female Holstein calves were housed in groups of 9, and all were fed 12 l/d milk and ad libitum grain starter and hay from automated feeders at grouping. Calves were allocated to three weaning strategies. (1) Early-Weaned (n=14): weaning began on d40, and milk allowance was gradually reduced until weaning was complete on d48. (2) Late-Weaned (n=14): weaning began on d 80 and was completed on d 89. (3) Weaned-by-Starter-Intake (n=28): weaning began when the calves consumed 200 g/d of starter, and was completed when the calves consumed 1,400 g/d. Each group pen contained 2 calves from each fixed-age treatment and 4 calves from the adjusted weaning age treatment. We recorded the daily quantities of milk, starter and hay eaten, and the frequency of rewarded and unrewarded visits of the feeder; we used unrewarded visit frequency as a sign of hunger. Body weights (BW) were recorded weekly and gains reported as % body weight per day. We estimated daily digestible energy (DE) intake based on milk and starter consumed. Analysis of variance was tested effects of group and treatment. For calves weaned according to starter intake, weaning began at 54.7±18.9 d (mean ± SD) of age, the duration of weaning was 21.1±10.6 d and ended at 75.8±10.7 d of age. During weaning, early weaned calves made more unrewarded visits to the feeder, than late weaned calves and calves weaned according to starter intake ($P=0.05$). Weight gain from week 3 to week 13 was significantly lower ($P<0.05$) for early weaned calves (1.56±0.08% BW) (mean ± SE) compared to both late weaned calves (1.85±0.09% BW) and calves weaned by starter intake (1.80±0.06% BW), with no difference between these last two groups ($P>0.10$). Three early weaned calves lost weight during weaning, whereas all late weaned calves and calves weaned according to starter intake gained weight. During weaning, early weaned calves ate less starter (0.45±0.18% BW) than either late weaned calves (1.41±0.18% BW; $P<0.001$) or calves weaned according to starter intake (1.11±0.12% BW; $P<0.001$). These differences in intakes resulted in early weaned calves having lower DE intakes (6.82±0.62% BW) than late weaned calves (8.63±0.62% BW; $P=0.04$) and calves weaned according to starter intakes (8.48±0.40% BW; $P=0.03$). Calves differ greatly in when they begin to eat solid feed. An advantage of automated feeders is that calves can be weaned at a variable age depending on their ability to eat solid feed, which reduces signs of hunger and improves weight gains during weaning.

Effect of two-stage weaning on behaviour and lying time in horses

Katrina Merkies[1], Kaitlyn Marshall[1], Severine Parois[2] and Derek Haley[3]
[1]University of Guelph, Department of Animal & Poultry Science, 50 Stone Rd East, Guelph N1E 2W1, Canada, [2]École Nationale Supérieure Agronomique de Rennes, 65 Rue de Saint-Brieuc, 35042 Rennes, France, [3]University of Guelph, Population Medicine, Ontario Veterinary College, Guelph N1E 2W1, Canada; kmerkies@uoguelph.ca

In natural conditions a foal gradually decreases nursing from its dam and continues to be in close physical proximity even past one year of age. With domestication, foals are typically separated both nutritionally and physically from their dams around six months of age. Past research has provided mixed results on short- and long-term effects of various artificial weaning methods. This study assessed the application of a physical barrier to prevent nursing prior to physical separation on the behavioural responses of mares and foals. Seventeen mares (mean±SEM; 9.8±4 years) and their foals (166±18 days) were assigned to either control (C) or a two-stage separation (TS) treatment whereby respective dams were fitted with an udder cover to prevent nursing for 4 d prior to physical separation. Control dams wore a sham udder cover, and foals nursed until physical separation, which occurred at the same time in both groups. All animals wore a triaxial accelerometer (Onset Pendant G) affixed to a hind leg for the duration of the experiment to record standing and lying behaviour every 20 sec. Instantaneous sampling every 10 min of behaviour (lying, locomotion [walking and running], aggression) was observed in-person for 8 h/day. Data was collected for 12 days: 4 days of baseline (BL: D1-4); 4 days with the udder covers or shams in place (PER 1: D5-8); and 4 days after complete physical separation of all mares and foals (PER 2: D9-12). Data was analyzed using a mixed procedure with repeated measures, with horse as a random variable. Sidak comparisons determined significant differences between treatments. Foals lay for a longer time and had more lying bouts than mares (34.91±12.4 min/bout/d in 8±3.4 bouts[a] vs 14.16±11.2 min/bout/d in 2±1.7 bouts[b] respectively; a,b differ $P<0.001$). During PER 1, C foals increased their total lying time compared to TS foals which did not differ from BL (32.9±2.4 min/bout/d in 10.8±0.5 bouts[a] vs 35.5±2.3 min/bout/d in 7.5±0.5 bouts[b] respectively; a,b differ $P<0.0031$). There was no treatment effect on lying in PER 2 ($P>0.75$). In mares there was no difference in lying duration or number of lying bouts across treatments ($P>0.99$), periods ($P>0.37$), or days ($P>0.10$). Physical separation increased the number of observations (obs) of locomotion in all animals (FOALS: 1.6±0.3[a] vs 7±0.8[b] avg obs; MARES: 1.1±0.3[a] vs 4.5±1.0[b] mean obs/horse/d, BL vs D9 respectively; a,b differ within group $P<0.001$), with C animals moving more than TS animals ($P<0.001$). More aggressive acts were observed in C than TS foals (3.9±0.5[a] vs 3.7±0.8[b] mean obs/horse/d, respectively during PER 2; a,b differ $P<0.01$) and more vocalizations were recorded upon separation in C than TS foals (33±16[a] vs 23±11[b] mean obs/horse/d, respectively on D9; $P<0.01$). Observations of aggression decreased in mares upon separation (5.4±1.6[a] (BL) vs 3.2±2.8[b] (PER 2) mean obs/horse/d; a,b differ $P<0.001$) with no treatment effect. In summary, two-stage weaning using udder covers decreased instances of vocalization and aggression in foals after separation from their dams, but did not appear to affect lying behaviour. Mares did not display any reduction in behavioural signs of stress with a two-stage approach to weaning. Two-stage weaning may mitigate behavioural stress responses in foals at weaning.

Effects of maternal vocalisations on the domestic chick stress response

Joanne Edgar, Suzanne Held, Ilana Kelland, Elizabeth Paul and Christine Nicol
University of Bristol, Clinical Veterinary Science, Langford House, Langford, BS40 5DU, United
Kingdom; j.edgar@bristol.ac.uk

We previously demonstrated that domestic hens are able to reduce their chicks' stress response. We also identified maternal vocalisations as a key component of the hens' response to chick stress; hens vocalise for around 10% of the time when their chicks are exposed to an air-puff, with some higher responding individuals calling for up to 33% of the time. We therefore sought to determine whether playback of maternal calls at 10% and 33% proportions would alleviate stress in non-brooded chicks. We also hypothesised that prior experience of the vocalisation might be necessary for this response. 72 chicks were assigned to one of two 'Experience' treatments, based on whether or not they would have prior experience of the maternal call playback. Then, during testing, behaviour and eye temperature responses of chick pairs were monitored during exposure to air-puffs at 30 second intervals during three time intervals (T1: 0-3 mins, T2: 4-6mins, T3: 7-9 mins). During testing chicks were split into 3 groups: Group 1 received no vocalisation playback (Control), Group 2 received playback of vocalisations for 10% of the time and Group 3 received playback of vocalisations for 33% of the time (n=12 pairs for each group). In response to the air-puff, chicks reduced sitting ($P=0.003$), pecking the environment ($P=0.002$) and increased freezing ($P=0.002$). They also showed a reduction in eye temperature ($P<0.001$). Although there were no behavioural differences between the three groups during air-puff exposure, Group 2 chicks showed a lower eye temperature response during T2 ($P=0.045$) and T3 ($P=0.030$) compared to Groups 1 and 2, suggestive of a stress-alleviating effect. There was no effect of prior experience on this response. Further research should focus on the stress-alleviating effects of additonal features of a broody hen.

The effects of group lactation sow housing on aggression and injuries in weaner pigs

Megan Verdon[1], Rebecca. S. Morrison[2] and Paul. H. Hemsworth[1]
[1]The Animal Welfare Science Centre, The University of Melbourne, Parkville, Vic, 3010, Australia,
[2]Rivalea Australia, Corowa, NSW, 2646, Australia; meganjverdon@gmail.com

We hypothesised that raising piglets in group-housed sow lactation systems will reduce aggression and injuries following mixing at weaning, in comparison to piglets raised in non-group lactation systems that were not mingled prior to weaning. Four treatments were applied to 72 sows and their litters (n=642 piglets). These were (1) Farrowing crate (FC; n=24 sows), (2) PigSAFE (PS; n=24 sows), (3) Farrowing crate and group lactation (GL_{FC}; n=12 sows), and (4) PigSAFE and group lactation (GL_{PS}; n=12 sows). FC and PS pigs remained in that environment from birth until weaning. GL_{FC} and GL_{PS} pigs were housed in FC and PS environments, respectively, from birth until 14 d of age after which they were transferred (with their dams) to group lactation pens (n=6 sows and litters/pen), where they remained until weaning. After weaning at 27 d, pigs were mixed into pens of four litters (average stocking density 0.54 m^2/pig) from the same treatment and behaviour recorded for 2 h. The following were calculated for each litter: frequency non-reciprocal (NR) aggression, fights (frequency, total duration, average duration, time to start fighting) and total aggression (fights + non-reciprocal aggression). Further, six piglets from each litter were randomly selected for skin lesion scoring on the day prior to weaning and 24 h post-mixing, so that lesions sustained at mixing could be calculated. Data were transformed where appropriate. Treatment effects were examined using a two-tailed univariate ANCOVA in conjunction with an LSD test. GL_{FC} and GL_{PS} pigs at weaning and mixing sustained fewer skin lesions (P<0.01), delivered less NR aggression (P<0.01), took longer to start fighting (P<0.01), fought less frequently (P<0.01), spent less time fighting (P<0.01), had shorter fights (P<0.01) and engaged is less total aggression (P<0.01) than either FC or PS pigs. These results emphasise the role of social experience in the development and regulation of aggression in the pig. Increasing knowledge of the factors that contribute to the development of pig aggression provides the opportunity to control the social environment and minimize the impacts of aggression on pig welfare and productivity. While group-lactation systems are thought to improve sow welfare through increased opportunity for movement and social interaction, this research provides evidence that they may also improve piglet welfare by reducing aggression and injuries at weaning. Further investigation is required to determine whether the benefits of socialisation are evident in the long-term.

Investigating sow posture changes: does environment play a role?

Gabriela M. Morello[1], Brian T. Richert[1], Donald C. Lay Jr.[2], Luiz H. A. Rodrigues[3] and Jeremy N. Marchant-Forde[2]
[1]Purdue Univ, Anim Sci Dept, 125 S Russell St, W. Lafayette, IN 47906, USA, [2]USDA-ARS, Livestock & Behav Res Unit, 125 S. Russell St, W. Lafayette, IN 47907, USA, [3]State Univ of Campinas (UNICAMP), Agric Engin Dept, Cidade Universitaria Zeferino Vaz, Campinas, SP, 13083-790, Brazil; gabymm@gmail.com

Alternative swine birthing systems have been designed and tested in an effort to replace conventional farrowing crates, due to welfare concerns. However, one major concern with all farrowing systems is the variation in pre-weaning mortality, mainly due to crushing by the sows. The reasons underlying the variation in sow maternal behaviors linked to crushing are not completely understood and many aspects of the sow's micro-environment within the farrowing room, and their possible effects on the sow, remain unknown. Therefore, the present study aimed to evaluate the impact of the environment on sow posture changes and on the incidence of crushing in farrowing crates within 48 hours post-partum. A total of 1,292 sows (Landrace × Large White, parity one through 10) and their litters were monitored through the course of one year (June 2013 – June 2014) in a commercial farrowing unit. Heated mat temperature (M) was obtained using an infrared thermometer. Ambient temperature (T), relative humidity (RH), light intensity (L), sound intensity (S) and air velocity (AV) were recorded using HOBO-U12 loggers and estimated hourly for each farrowing crate. Behavior was continuously recorded for a total of 30 sows within crates of distinct thermal conditions. Procedure GLMSELECT (stepwise) was used on SAS to select the model which most explained the variation in posture changes. Crushing accounted for approximately 63% of all mortality (average 0.8 ± 1.2 piglets crushed/sow, range 0-9). Preliminary data analysis showed that sows that spent 88% of the 48 h post-partum period exposed to T above 24 °C and at least 3% of this time above 26 °C lay down faster (5.3 ± 1.2 vs 9.6 ± 0.4 s, $P=0.06$) than sows not exposed to temperatures above 26 °C. Sows exposed to T above 26 °C showed a reduction ($P=0.05$) in the total posture changes per hour and an increase ($P=0.02$) in total time spent lying laterally. In crates where heat mats had temperatures above 32 °C, sows lay down quicker than sows in crates with lower heat mat temperatures (8.4 ± 0.3 vs 12.8 ± 0.7 s $P<0.01$), possibly due the behavioral interaction between sows and their piglets. The preliminary results indicated that changes in the thermal environment of the farrowing crates impact behaviors that are relevant for piglet crushing. The effects of AV, RH, L and S on the behavior of sows are currently under investigation. A data mining classification technique is being used to determine the implications of these changes on piglet crushing by the sows.

Effects of floor space allowance on aggression and stress in grouped sows

Paul Hemsworth[1], Maxine Rice[1], Rebecca Morrison[2], Kym Butler[3] and Alan Tilbrook[4]
[1]Animal Welfare Science Centre, The University of Melbourne, Parkville, 3010, VIC, Australia, [2]Rivalea Australia, Corowa, 2646, VIC, Australia, [3]Biometrics Group, Department of Economic Development, Jobs, Transport & Resources, Parkville, 3010, VIC, Australia, [4]Animal Welfare Science Centre, South Australia Research and Development Institute, Roseworthy, 5371, SA, Australia; phh@unimelb.edu.au

A recent experiment found that increasing floor space from 1.4-3 m^2/sow was associated with declines in aggression and stress early after mixing, and increases in farrowing rate. While the results were in accord with a linear decline in cortisol and aggression at day 2 after mixing from 1.4-3 m^2/sow, the results were also in accord with a decline in cortisol and aggression, and no further decline above 1.8 m^2/sow. There was insufficient precision to determine which scenario is more biologically correct due to spatial variability in aggression and stress within the research facility. To minimise spatial variability within the same facility in the present experiment, similar-sized pens with varying groups sizes (10-20) in 4 separate blocks of 3 contiguous pens within each of 9 time replicates (180 sows/replicate) were used to examine 6 space treatments (1.45-2.9 m^2/sow) with treatments randomised to pens. While it may be argued that space allowance is confounded with group size in this design, in our previous experiment there were no group size effects, for pens of 10-80 sows, or appreciable interactions between space and group size on aggression, stress and reproduction. Sows were introduced to treatments within 4 days of insemination and were floor-fed 4 times per day (2.5 kg/sow/day). On days 2 and 26 post-mixing, aggression at feeding and plasma cortisol concentrations were measured. REML mixed model analyses examined treatment effect after accounting for replicate and random spatial location effects within replicate. There was a consistent linear effect of space on aggression ($P<0.0001$) and plasma cortisol ($P<0.0001$) at day 2, with aggression and stress declining with increasing space. There were no effects of space at day 26 ($P=0.14$ and $P=0.79$, respectively). These results show that increased floor space immediately post-mixing reduces aggression and stress and that sows may adapt to reduced floor space over time. While floor feeding is competitive, accessing feeding stalls or electronic sow feeder stalls also leads to competition between group-housed sows, and thus the effect of space on stress in the present experiment has implications for group housing in general. Staged-gestation penning, with more space immediately after mixing and less space later in gestation, may address both animal welfare and economic considerations.

Lameness and claw lesions in group housed sows: effect of rubber-topped floors

Emilie-Julie Bos[1,2], Dominiek Maes[1], Miriam Van Riet[1,2], Sam Millet[2], Bart Ampe[2], Geert Janssens[1] and Frank Tuyttens[1,2]
[1]*Ghent University, Faculty of Veterinary Medecine, Salisburylaan 133, 9820 Merelbeke, Belgium,* [2]*Institute for Agricultural and Fisheries Research (ILVO), Animal Sciences Unit, Scheldeweg 68, 9090 Melle, Belgium; emiliejulie.bos@ilvo.vlaanderen.be*

Lameness in sows is a major welfare and production problem and it is often related to claw problems. Prevalence of lameness and claw lesions caused by inadequate pen floor quality is increasing with compulsory group housing for pregnant sows. Generally sows are housed on concrete solid or slatted floors. In a 2×3 factorial design, we investigated the effect of rubber top layers on concrete floors in the group pens and the effect of three levels of zinc supplementation (0, 50, 100 ppm) in the feed on gait and claw lesions, as previous studies have shown a positive effect of zinc supplementation on claw quality and horn production. Therefore, 120 hybrid sows (6 groups of 20) were followed, in a longitudinal study over three reproductive cycles. During gestation (day 28-108 of the cycle) three groups were housed on concrete floors (40.3 m^2 slatted and 31.7 m^2 solid), and three groups in pens with all the slats and half of the solid floors covered with rubber (40.3 m^2 slatted and 15.8 m^2 solid). Each cycle, we observed sows' gait five times and claws (eight parameters per claw) two times on a continuous scale (tVAS) of 150 and 160 mm, respectively. The data were analysed using a linear mixed model with, floor, feed, phase in cycle and their interactions as fixed effects and parity, group and sow as random effect. As there was no significant interaction of feed × floor for the observed parameters, this abstract focusses on floor, gait and claw lesions. At the end of gestation, sows housed in pens provided with a rubber-topped floor scored 6.6% better on gait compared to sows in pens fully in concrete (P=0.01) (9.88±4.12 mm on tVAS). Floor affected claw lesions during mid-gestation and at the end of gestation. At mid gestation, both heel overgrowth and erosion (4.6±1.8 mm; P=0.01) and heel-sole crack (3.1±1.5 mm; P=0.04) scores were higher (i.e. worse) on floors with rubber top layer. However, vertical cracks in wall horn (3.4±1.7 mm; P=0.04) scored higher on concrete floors. At the end of gestation, both white line (2.9±1 mm; P=0.02) and claw length (4.7±1.4 mm; P<0.001) had higher (i.e. worse) scores on rubber-topped floors. Although rubber-topped floors did not result in a consistent improvement of all types of claw problems, improved gait scores at the end of gestation suggest that this floor type has a positive effect on welfare.

Promoting positive welfare through environmental enrichment

Ruth C. Newberry
Norwegian University of Life Sciences, Department of Animal and Aquacultural Sciences,
Arboretveien 6, P.O. Box 5003, 1432 Ås, Norway; ruth.newberry@nmbu.no

In 1995, I defined environmental enrichment as an improvement in the biological functioning of captive animals resulting from modifications to their environment. Focus was placed on easily measurable physical health outcomes, emphasizing that, by reducing the worst negative impacts of barren housing systems, enrichment could have practical benefits for animal husbandry. Since then, environmental enrichment has become a standard procedure incorporated into many animal welfare regulations and guidelines. Provision of manipulanda to redirect harmful behaviour such as tail biting remains the primary approach in practice. Enrichments are often minimal and their effectiveness modest due to continued constraints relating to manure handling, space and labour costs, and concerns about animal health risks. It is time to raise the bar. Animal welfare has been conceptualized has having three dimensions, relating to physical body condition, psychological feeling states, and naturalness of behaviour. These dimensions can be viewed as contributing to biological fitness at different organizational levels, from bodily responses to current conditions, through short-term anticipatory responses, to long-term adaptability and stress resilience. Scientific advances now enable us to assess animal welfare at all three levels of response, allowing for a more refined approach to environmental enrichment than reducing injury and death rates. Thus, we are in a position to implement enrichment methods that produce positive welfare outcomes at all three response levels, including body condition correlates of reproductive fitness, positive anticipation, and behavioural flexibility. To achieve this goal, enrichments need to be both diverse and dynamic, accommodating individual differences and developmental changes arising from habituation, neurogenesis and 'learning to learn.' Although challenging, we need to move away from a 'one size fits all' mentality and allow animals greater behavioural freedom of expression. Only then shall we have fully embraced the concept of environmental enrichment.

Environmental enrichment in kennelled Pit Bull Terriers (*Canis lupus familiaris*)

Jenna Kiddie and Anna Bodymore
Anglia Ruskin University, Life Sciences, East Road, Cambridge, CB1 1PT, United Kingdom;
jenna.kiddie@anglia.ac.uk

Although dogs rescued from dog-fighting rings may no longer be exposed to aversive experiences associated with dog-fighting they are likely to still experience negative emotional states associated with kennelling. Environmental enrichment can help dogs cope with the kennel environment. Although social enrichment can be beneficial it places a larger demand on the carers and may not be appropriate in under-resourced kennels. This study therefore aimed to assess the effects of easily accessible physical environmental enrichment on the behaviour of kennelled Pit Bull Terrier types dogs rescued from a dog-fighting ring within the Philippines. Thirty-six dogs were pseudo-randomly recruited from a rescue shelter and allocated to one of three treatment groups following a matched-subject design: (1) cardboard bed provision; (2) coconut provision; and (3) visual contact with dogs housed in adjacent cages obstructed with cardboard partitions. Behavioural diversity was calculated from activity budgets calculated from behavioural data collected over three phases: baseline, enrichment and post-enrichment. Behavioural diversity and individual behaviours were analysed using linear mixed-effect models, with treatment group and sex fitted as fixed-effect factors, dog fitted as a random factor, and day and phase fitted as repeated measures. Only one test (drinking frequency) showed a significant effect. Treatment had no effect, but females were found to drink more than males ($F_{1, 207}=10.423$, $P=0.001$). Subject also had an effect on drinking ($Z=1.965$, $P=0.049$) as did the interaction between phase and day ($P<0.01$ in all combinations). Variables were further analysed with phase fitted as a fixed-effect factor and dog as a random factor. Four behaviours (lying, sitting, yawning and other) differed significantly between phases, with varying degrees of individual variation. In conclusion, enrichment, regardless of type, had some effect on the dogs, with many of the effects depending on the sex and the individual.

Do farmed American mink (*Neovison vison*) prefer to eat using more naturalistic postures?

María Díez-León, Margaret Quinton and Georgia Mason
University of Guelph, Department of Animal and Poultry Science, 50 Stone Road East, Guelph ON N1G 2W1, Canada; mdiez@uoguelph.ca

Cage size regulations typically assume that taller cages are better for animal welfare, since providing more space should increase abilities to display more diverse postures (e.g. standing upright) and natural behaviours. New regulations in North America therefore recommend increasing ceiling heights for mink cages to at least 30 cm, while in Europe cages must be 53 cm high. However, mink are fed by placing feed on the cage top. Because small, young mink must thus climb to feed, and even as adults, this practice requires them to stand to feed. This is not something they would do in the wild: naturally, mink typically stand over their food, eating prey on the ground. We therefore investigated whether this aspect of husbandry matters for mink welfare, testing the hypotheses that: (1) young mink prefer to feed from lower heights; and (2) these early feeding preferences wane as they grow in size. We housed 64 independent male-female pairs at 2.5 months old in enriched cages measuring 75L×61W. Most of the cage measured 46 cm in height and we modified each cage's feeding strip (15L×61W) so that it would accommodate four different heights to feed from: 25, 38, 46 or 53 cm above the floor (spanning the range of cage heights used on North American and European farms). The order these heights were presented in varied across cages to avoid any systematic confounds. Mink were habituated to feed from each of the four heights for four weeks (each mink feeding from each height at least twice). Food was then delivered to all four heights simultaneously and the mink observed, a procedure that was repeated once a month when subjects were 3, 4, 5, 6 and 7 months old. Preference was assessed by calculating the proportion of feeding behaviour allocated to each height, and assessing the percentage of food left uneaten – both on these days when mink had free choice, and also on other days when they could only feed from one height. Preferences to feed from each height varied with age and sex ($F_{3,490}$=3.80; $P<0.0001$). Females always preferred to feed from the lowest height ($P<0.01$ for all significant contrasts). For males, although they always avoided the highest height, the preference for the lowest height only emerged at 5 months old ($P<0.01$ for all significant contrasts). Preliminary results on food leftovers are consistent with a preference for feeding at lower heights. This suggests that mink prefer to feed in ways that are less dissimilar to their natural feeding posture, not just as juveniles, but even when they have reached adult size. Future research will investigate cage height preference for behaviours other than feeding. This has important implications for developing animal welfare regulations regarding cage dimensions.

Housing implications on the behavioural development of red fox (*Vulpes vulpes*) cubs

Sandra Alvarez Betancourt, Innes Cuthill and Stephen Harris
University of Bristol, Biological sciences, University of Bristol. Life Sciences Building. 24 Tyndall Avenue., BS8 1TQ Bristol, United Kingdom; bzsab@bristol.ac.uk

We aimed (1) to determine if social status of captive red fox (*Vulpes vulpes*) cubs can be predicted at an early stage; (2) to identify the main behaviours involved in the establishment of an intra-litter hierarchy, and the time at which it is established; and, (3) by looking at stereotypic and aggressive behaviour, to determine the relationship between status and welfare of cubs in a captive environment. This information can be applied to management and captivity programs where negative behaviours such as stereotypy can then be addressed individually from an early stage. Twelve litters reared at wildlife hospitals were recorded continuously for an average of 70 days. Cubs were scan sampled every 20 minutes and 42 behaviours were recorded following a previously created ethogram, with a particular focus on social behaviours. Activity budgets and sleeping proximity of the cubs was recorded. Linear discriminant analyses (LDA) were used to identify which behaviours are the best predictors of social status within and between litters. Frequencies and duration of fights, stereotypic behaviours, duration of feeding events and average sleeping proximity distance were compared within different enclosure characteristics (size, number of nests, presence of toys) using chi-square tests. Activity budgets were compared using the Kappa coefficient of agreement. Three main stages of behavioural development were identified: (1) period of low activity (0-3 weeks old) and low synchrony between cubs (kappa=0.46). LDA results showed that sucking behaviour was the main difference between littermates during this stage; (2) period of high aggressiveness and activity (4-10 weeks old) when synchrony increased (kappa=0.64). Groom, social play, bite, pounce, stand over, muzzle, stereotypy, wrestle, fight for food and running alone were the main discriminants between litter-mates according to the LDA. From these, the frequency of agonistic behaviours was on average higher in dominant cubs. (3) Period of moderate activity and high amicable socialization (11 weeks on) and high synchrony (kappa=0.85). Mount, groom, bite, bare teeth, wrestle, scent marking and stereotypy were the main determinants of intra-litter variance. Frequency of fights and stereotypy were proportional to the number of cubs in the litter and inversely proportional to enclosure size. Sleeping proximity was inversely proportional to enclosure size. In conclusion, social status of cubs can be predicted by early behaviours. Cubs at both ends of the hierarchy present higher aggressiveness and stereotypy, and these behaviours are negatively correlated with enclosure size and number of external stimuli.

Normal behaviour of housed dairy cattle with and without pasture access: a review

S. Mark Rutter

Harper Adams University, Animal Production, Welfare and Veterinary Sciences, Edgmond, Newport, TF10 8NB, United Kingdom; smrutter@harper-adams.ac.uk

Recent research at Harper Adams University and elsewhere into dairy cow preference for pasture versus cubicle housing is providing a new insight into what constitutes 'normal behaviour' for housed cows. This review discusses the implications of this research for dairy cow management and welfare. In bad weather and the winter, cows preferred the shelter provided by cubicle housing. Previous experience had a big effect on pasture preference: the longer calves/heifers/cows were reared without experience of pasture the stronger their preference for cubicle housing. The ontogeny of grazing also required pasture experience i.e. the innate foraging behaviour of calves is to suckle and calves/heifers reared without pasture access have to learn to graze. These results raise the question: if cattle are to be housed for part of the year, would it be better to house them continuously? Other results suggest not, as there are clear production, health and welfare benefits to pasture access. Cows at pasture had lower levels of lameness and mastitis. Cows with pasture access also produced more milk than those continuously housed, and produced less methane per kg milk. About half of this extra milk was attributed to grass intake, and increased lying, improved comfort and/or lower stress probably accounted for the rest. Incorporating free access to pasture is difficult on many farms, and confining high-yielding cows at pasture for long periods without access to a Total Mixed Ration compromised production and welfare. Developments in precision livestock farming offer the potential to provide a technological solution to this problem. These research findings could be used as the basis to design novel, adaptive housing that responds to cow behaviour. The aim would be to incorporate the best aspects of pasture with the best aspects of housing to provide an environment that meets the needs of cows all year around.

Effect of outdoor access and shelter type on broiler behaviour, welfare, and performance

Lisanne Stadig[1,2], T. Bas Rodenburg[3], Bart Ampe[2], Bert Reubens[2] and Frank Tuyttens[1,2]
[1]Ghent University, Salisburylaan 133, 9820 Merelbeke, Belgium, [2]Institute for Agricultural and Fisheries Research (ILVO), Burg. v. Gansberghelaan 92, 9820 Merelbeke, Belgium, [3]Wageningen University, De Elst 1, 6708 WD, the Netherlands; lisanne.stadig@ilvo.vlaanderen.be

Outdoor access gives broilers more space and opportunities to perform natural behaviour, and is assumed to improve their welfare. However, the effect on specific welfare measures and performance is unclear, and chickens rarely make optimal use of the outdoor area. Shelter can encourage chickens to use the outdoor more frequently and homogeneously, but the ideal type of shelter is still unknown. Dense vegetation could be more attractive to the birds than artificial shelters. The aim of this study was to investigate behaviour, welfare and performance of chickens without outdoor access, with access to grassland with artificial shelters, or with access to dense woody vegetation. Slow-growing broilers (Sasso T451; n=600; 2 rounds) were housed indoor from d0-70 (IN, n=4 groups of 50), or given access either to grassland with artificial shelters (triangular wooden shelters; GR, n=4 groups of 50), or to willows (WI, n=4 groups of 50) from d21. Feed provision and weight were recorded at group level. For the outdoor groups, number of birds inside was recorded hourly (7 am-9 pm). In week 10, tonic immobility (TI) tests were performed on 5 birds per group, and foot pad and hock dermatitis, lameness and cleanliness were scored for 10 birds per group, according to the Welfare Quality® (WQ) method. WQ scores were dichotomized for proper analysis, converting scores >1 to 1. Data were analysed using proc mixed and proc glimmix in SAS, correcting for repeated measures by date, and stable nested in round. Housing type did not affect TI duration, lameness or foot pad dermatitis. Weight of IN birds at d70 was higher than that of GR and WI birds (LS means±se: 2.79±0.02 vs 2.66±0.02 and 2.68±0.02 kg, P=0.005). However, mean feed intake did not differ. Prevalence of hock dermatitis was higher for IN than for GR (39 vs 7% with dichotomized score 1; P=0.007), and tended to be higher than for WI (13%, P=0.07). IN birds tended to be dirtier than GR birds (89 vs 72% with dichotomized score 1; P=0.07). Mean outdoor use was higher in WI than in GR chickens (means: 47.9 vs 33.0%; P<0.001). Time of day also influenced outdoor use, with most chickens being outside between 7 and 9 am and between 6 and 9 pm (P<0.001). Concluding, outdoor access had a slightly negative impact on growth efficiency but a positive effect on some welfare indicators (hock dermatitis, cleanliness). The chickens ranged more in outdoor areas with willows than in grassland with artificial shelters, indicating this type of shelter to be more suitable.

Effect of enrichment material on hair cortisol and chromogranin A in pigs

Nicolau Casal[1,2], Xavier Manteca[1], Damián Escribano[3], José Joaquín Cerón[3] and Emma Fàbrega[2]
[1]Universitat Autònoma de Barcelona, School of Veterinary Science, 08193 Bellaterra, Spain, [2]IRTA, Veïnat de Sies s/n, 17121 Monells, Spain, [3]Universidad de Murcia, School of Veterinary Medicine, 30003, Murcia, Spain; nicolau.casal@irta.cat

Cortisol (C) concentration in hair and to a lesser extent chromogranin A (CgA) in saliva, have been suggested as non-invasive indicators of chronic stress. The aim of this experiment was to determine whether enrichment material (EM) can reduce hair cortisol and CgA salivary levels in pigs. EM was tested in a total of 56 male growing pigs [(Landrace × Large-white) × Pietrain] randomly divided into 2 groups at the age of 15 weeks; one group of 28 animals was provided with enrichment material (EM+) consisting of sawdust, ropes and rubber balls and the other remained as control (EM-). Three samples of hair and saliva were taken just before the provision of EM and one and two months after the start of the treatment. Repeated measure analyses (PROC MIXED for C and GENMOD for CgA) were done. Correlations between C and CgA were analysed using a Parametric Pearson's rank correlation test. No differences before the provision of EM were found neither in Cortisol (11.93 vs 12.34 pg/mg) nor CgA (1.14 vs 1.16 µg/ml), between the subsequently EM- and EM+ groups, respectively. One month after starting the treatment, both cortisol and CgA were significantly lower ($P<0.001$) for the two groups compared to the initial levels; CgA concentration was also lower in EM+ compared to EM- (0.36 vs 0.82 µg/ml, $P<0.001$) but no differences were found between EM+ and EM- for cortisol levels (5.71 vs 6.78 pg/mg, $P=0.883$). After two months, EM+ animals had lower levels compared to EM- for both CgA (0.26 vs 0.54 µg/ml, $P=0.009$) and C (3.96 vs 7.58 pg/mg, $P<0.001$). Furthermore, both CgA and C were lower compared to the initial levels ($P<0.001$). After two months, CgA levels were also lower than after one month ($P=0.003$) in EM- pigs ($P=0.002$) but not in EM+. For the same period, cortisol levels also showed a tendency tended to decrease ($P=0.063$), but in this case, differences were found for EM+ ($P=0.011$) but not for EM-. Significant correlations (r=0.478, $P<0.001$) were found between CgA and C values. In conclusion, these results suggest that enrichment material can reduce the levels of hair cortisol and salivary chromogranin A in pigs and that hair cortisol and chromogranin A can be a useful tool to detect chronic stress.

Ammonia on a live export shipment, effects on sheep behaviour and development of an effective ammonia sampling strategy

Yu Zhang, Mat Pines and Clive Phillips
Centre for Animal Welfare and Ethics, School of Veterinary Science, University of Queensland, Building 8134, White House, CAWE, Gatton Campus, University of Queensland, QLD 4343, Australia; yu.zhang2014@gmail.com

The microclimate in a ship can potentially affect the health and welfare of exported sheep. Adverse effects of ammonia accumulation on sheep welfare have been monitored in simulated live export conditions but never tested on ship. It is essential to monitor gaseous ammonia concentrations, especially in the pens with poor ventilation, high temperature and humidity. However, ammonia measurement on ships is currently unregulated in Australia and little is known on the within-pen gaseous ammonia sampling scheme during live export. An experiment was firstly conducted during a live export shipment to investigate the behaviour responses of sheep to high and transient ammonia concentrations in pens with insufficient ventilation, high temperatures and humidity on a voyage with sheep from Australia to Middle East. Sheep exposed to high and transient concentrations of ammonia stood longer and spent less time feeding and ruminating, held their head higher, probably to avoid the greater ammonia concentrations near the floor, and had more conjunctivitis. Subsequently, we evaluated different sampling schemes by estimating the margin of error of the microclimate data on the same voyage. We synthesized and analysed the data on gaseous ammonia, and dry-bulb temperature and relative humidity in 18 pens, which were selected as 9 high and 9 low ammonia concentrations, and 16 locations within each pen. The number of measured pens contributed more to the variance of microclimatic measurements than the number of sampling locations within each pen. It is recommended that at least 10 pens (5 pens with predicted high versus low ammonia concentration, respectively) with 2 locations in each pen should be sampled randomly during the voyage. This study highlights the importance of a suitable sampling strategy to detect ammonia, which has adverse effects on sheep behaviour on board ship.

Effects of bedding with recycled sand on lying behaviours of late-lactation Holstein dairy cows

Heather DeAnna Ingle, Randi A. Black, Nicole L. Eberhart and Peter D. Krawczel
The University of Tennessee, Department of Animal Science, 2506 River Drive, 258 Brehm Animal Science, Knoxville, TN 37996-4574, USA; krawczel@utk.edu

Sand bedding can promote the overall welfare of freestall-housed dairy cattle, but presents a variety of management issues. One approach to managing sand bedding to alleviate some of the negatives is removing manure and displaced sand by flushing with water and separating the sand within a settling lane. However, there is an increased moisture and organic matter, relative to unused sand, that results from this management strategy. As previously demonstrated with organic bedding materials, increased moisture can affect how cows use their freestalls. Establishing the behavioural response to recycled sand from dairy cows was the first step in evaluating its suitability as bedding. Our objective was to determine the effects of bedding with recycled sand on lying behaviours and activity of late-lactation Holstein dairy cows. Cows (n=32) were divided into four groups (8 cows per group), which were balanced by DIM (265.5±34.1 d). Late-lactation cows were used to minimize the long-term effects on udder health and productivity, if the recycled sand increased the incidence of mastitis. Treatments were bedding with recycled sand (RS; reclaimed from the dairy's flushing system) or control (CO; clean, dry sand) and cows were exposed to each using a cross-over design with 7-d periods from Jan to Feb, 2015. Throughout the study, bedding was kept at curb height and freestall stocking density at 100%. Accelerometers, attached to the rear legs, recorded daily lying time (h/d), lying bouts (n/d), lying bout duration (min/bout), and steps (n/d). Cows were habituated to pens and dataloggers for 2 d prior to data collection. A mixed model analysis was used to determine effects of treatment and treatment sequence (SAS v9.3). Sequence was not significant for any response variable ($P \geq 0.17$). Treatment did not affect lying time ($P=0.4$; mean for RS and CO was 11.7±0.4 h/d), lying bouts ($P=0.89$; mean for RS and CO was 8.8±0.5 per d), or lying bout duration ($P=0.83$; mean for RS was 86.6±4 min/d and CO was 86.6±4 min/d). CO cows took more steps than RS cows (1,728±103 vs 1,570±102 n/d, respectively; $P=0.02$). Recycled sand had little to no effect on the lying behaviour of late lactation Holstein dairy cows during winter months. This suggests, from a behavioural perspective, recycled sand may be a viable bedding material for dairy cows. However, the effect of recycled sand on udder hygiene, udder health, and milk quality must be established before it can be recommended as a management practice.

Cooling dairy cows efficiently with water: effects of sprinkler flow rate on behavior and body temperature

Jennifer M. Chen[1], Karin E. Schütz[2] and Cassandra B. Tucker[1]
[1]*University of California, Davis, CA, USA,* [2]*AgResearch, Ltd, Hamilton, New Zealand;*
jmchen@ucdavis.edu

Sprinklers reduce heat load in cattle, but elicit variable behavioral responses: cattle readily use this resource in some studies, but avoid wetting the head or entire body in others. Some of this behavioral variation may be explained by sprinkler attributes – such as flow rate – that differ across studies. Our objectives were to determine how flow rate affects dairy cattle behavior and body temperature (BT) and to evaluate cooling effectiveness against water use. We administered 3 treatments at a shaded feed bunk at the U.C. Davis dairy facility: an unsprayed control and 1.3 or 4.9 l/min sprinklers (both used commercially; 3 min on, 9 min off, 24 h/d). Data were collected from pairs (n=9) of lactating Holsteins in summer [24-h maximum air temperature (T) = 33±3 °C, mean ± SD]. Each pair of cows received 1 treatment/d for 2 d each, with order of exposure balanced in a crossover design. BT was recorded at 2-min intervals with loggers. Visits to the feed bunk and head posture when walking through spray were measured continuously from video recordings. BT and visits to the feed bunk did not differ for the 2 flow rates (GLM; $P \geq 0.152$); thus, all differences are described between the sprinkler treatments and the control. Sprinklers did not affect how much time cows spent at the feed bunk overall (mean ± SD: 5.7±1.0 h/d; GLM; $P=0.458$), but changed how they used this area: when sprinklers were present cows visited the feed bunk $\geq 13\%$ less often than in the unsprayed control (GLM; $P \leq 0.036$), but each bout was $\geq 23\%$ longer (GLM; $P \leq 0.001$). This change in bout structure may reflect cows' reluctance to wet their heads, as moving toward and away from the feed bunk sometimes meant walking through spray. When the sprinklers were on, cows left the feed bunk about half as often as expected by chance (1-sample t-test; $P \leq 0.010$), and they lowered their heads ≥ 1.8 times more often when walking through spray than when there were no sprinklers (GLM; $P \leq 0.024$). Sprinkler use increased with heat load: cows spent more time at the feed bunk with sprinklers when the weather was warmer (≥ 22 min/d per 1 °C T increase; GLM; $P<0.001$). With sprinklers, BT between 1300 to 2,00 h was ≥ 0.3 °C lower than when there was only shade (GLM; $P \leq 0.015$). In summary, although cows avoided wetting their heads, they willingly used feed bunks with sprinklers, likely for the cooling benefits received. When there were sprinklers, BT was lower than when there was only shade, and cows spent more time at the feed bunk as T increased. The 2 flow rates had similar effects on cattle behavior and BT, despite differing nearly 4-fold in the amount of water used; this suggests an opportunity for reducing water usage.

Breeding but not management change may simultaneously reduce pig aggressiveness at regrouping and in stable social groups

Simon P Turner, Suzanne Desire, Richard B D'eath and Rainer Roehe
SRUC, Edinburgh, EH9 3JG, United Kingdom; simon.turner@sruc.ac.uk

Aggression between pigs is routine at regrouping and in stable social groups and needs better control. We examine if the number of skin lesions from regrouping and chronic aggression are associated to inform whether efforts to reduce aggression in both contexts may be complementary or antagonistic. Phenotypically, individuals (n=1,166) or social groups (n=78) that fight more at regrouping, even if defeated, show few injuries from chronic aggression (group level, r=-0.28 to -0.38, $P<0.05$). Therefore regrouping aggressiveness minimises future chronic aggression making it difficult for management to reduce both forms of aggression simultaneously. However, some pigs show few injuries from both regrouping and chronic aggression. Understanding their behaviour may inform management strategies that increase these desirable outcomes. Cluster analysis grouped pigs (n=1,163) with >80% similarity in regrouping aggressive behaviour to identify characteristics of pigs with few lesions from both regrouping and chronic aggression. Although pigs clustered into behaviourally distinct groups, those with high similarity in behaviour showed divergent amounts of skin lesions, suggesting that there was no single 'low lesion' strategy that management should target. At the genetic level, delivery of aggressive behaviour at regrouping is heritable ($h^2>0.31$ s.e. 0.04) as is the lesion count from regrouping or chronic aggression ($h^2=0.19$ s.e. 0.02 to 0.43 s.e. 0.04). The lesion count is genetically correlated with reciprocal fighting and therefore indicates aggressive propensity (maximum $r_g=0.80$ s.e. 0.05). Lesion counts at 24 h and 3 wks post-regrouping are positively genetically correlated (r_g 0.28 s.e. 0.07-0.50 s.e. 0.09) suggesting that breeding could simultaneously reduce regrouping and chronic aggression. These findings suggest that phenotypically suppressing regrouping aggression through management change alone may be counter-productive, as chronic aggression may increase. Although some pigs receive little aggression in both situations, they do so by a variety of strategies. Breeding, however, has greater potential to simultaneously reduce regrouping and chronic aggression.

Effect of social stressors on behaviour and faecal glucocorticoid in pregnant ewes

Nur Nadiah Md Yusof[1,2]*, Jo Donbavand*[2]*, Susan Jarvis*[2]*, Kenny Rutherford*[2]*, Leonor Valente*[2,3] *and Cathy Dwyer*[2]
[1]*University of Edinburgh, School of Biological Sciences, Darwin Building, Max Born Cresecent, EH9 3BF Edinburgh, United Kingdom,* [2]*Scotland's Rural College, West Mains Road, EH9 3JG Edinburgh, United Kingdom,* [3]*University of Nottingham, School of Bioscience, Sutton Bonington, LE12 5RD Leicestershire, United Kingdom; nadiah.yusof@sruc.ac.uk*

Stress experienced during pregnancy can compromise offspring development, and adversely affect the pregnant mother. The aim of this experiment was to investigate whether aspects of normal farm management cause stress in pregnant ewes. Seventy-seven twin-bearing crossbred ewes were divided into 6 control (C) and 5 stress (S) treatment groups (7 ewes per group), balancing for parity: primiparous (P) and multiparous (M). From weeks 11-18 gestation, S treatment ewes had smaller space allowance (1.18 vs 2.7 m^2) and feeding trough space per ewe (36 vs 72 cm) compared to C ewes, and underwent two social mixing events. Aggressive behaviour at feeding, and ewe behavioural time budgets were recorded at 12, 14, 16 and 18 weeks of gestation while faecal samples were collected on 4 occasions. Data were analysed by Residual Maximum Likelihood procedure (REML) in Genstat 16. S ewes displayed significantly higher aggressive behaviour at week 16 of gestation (mean frequency: S=138 vs C=92; $P<0.01$). At 18 week of gestation, M ewes were more aggressive than P when feeding regardless of treatment ($P<0.01$). P ewes spent significantly more time ruminating than M ewes at week 16 of pregnancy ($P=67.6\%$ vs M=32.4%; $P<0.01$). S ewes spent more time ruminating at week 18 of pregnancy compared to C ewes (62% vs 38% respectively; $P<0.05$). Four days after allocation to treatment, S ewes tended to have higher faecal glucocorticoid concentrations than C ewes ($P=0.07$). P ewes had significantly higher concentrations of faecal glucocorticoids than M ewes at week 12 ($P<0.001$), 14 ($P<0.001$) and 16 ($P=0.005$) of gestation. The study suggests reduced pen and feeder space caused an increase in aggression and ruminating, and may have induced stress. However, maternal experience had a greater effect on glucocorticoid level than treatment. This may be related to a greater metabolic effort required in first pregnancies.

Effects of mining noise frequencies and amplitudes on behavior, physiology and welfare of wild mice (*Mus musculus*)

Karen Mancera[1], Peter Murray[2], Marie Besson[1,3] and Clive Phillips[1]
[1]Univeristy Of Queensland, Centre of Animal Welfare and Ethics, Warrego Hwy 5391, Bld 8134, UQ-Gatton Campus, 4343, Gatton, Australia, [2]University of Queensland, School of Agricultural and Food Sciences, Warrego hwy, 5391, Bld 8150, UQ-Gatton Campus, 4343, Gatton, Australia, [3]École Nationale Supérieure Agronomique de Rennes, 65 Rue de Saint-Brieuc, 35042 Rennes, France; dra.kelokumpu@gmail.com

We exposed wild mice (*Mus musculus*) to a simulated mining noise sequence containing open cast mining machinery. To investigate posiible damaging effects, two levels of amplitude were set to exemplify energy inputs encountered at 0-500 meters (70-75 dB; High Amplitude treatment (HA)) or 500- 1000 meters (60-65 dB; Low Amplitude Treatment (LA)) from the mining site. In a second experiment, two isoenergetic levels of frequency at HA volume were explored: Low Frequencies ($\leq$2,000 Hz, LF) and High Frequencies (>2,000 Hz, HF). A Control (C), where animals were exposed to only background laboratory noise below 55 dB was included in both experiments. Mice were exposed to treatments for 3 weeks and behaviour was recorded continuously to analyze stress-related behaviours. Fecal samples were collected daily for corticosterone analysis. Animals were euthanized as part of the colony's procedures and spleen, adrenal glands and thymus were collected, weighed and preserved for analysis. General linear models were used for statistical analysis. In the amplitude experiment there was an increase in the time spent circling (stereotypical behaviour) in the HA treatment compared with LA and C (HA=34.7 s/h, LA=8.1 s/h, C=3.4 s/h; $P<0.001$). In addition, fecal corticosterone levels were increased only for LA when compared with C, with HA intermediate ((LA=251.9 ng/ml, HA=229.09 ng/ml, C 208.93 ng/ml=; P=0.003). Thus, evidence suggests amplitude-dependent stress responses where behavioural stereotypies downregulated systemic corticosterone at high amplitudes. Also, HA males' spleens were lighter than C, suggesting adverse effects on immunocompetence (HNA=0.017 g; C=0.023 g; P=0.042). The adrenal cortex was reduced in HA females compared to C (P=0.036), an alteration previously seen in response to acute stress. Regarding frequencies, HF females had lighter spleens compared with LF and C (HF=0.023 g, LF=0.030 g, C=0.032 g, P=0.049) while other histological variables were non-significant. In conclusion, mining noise adversely affected the behavior and welfare of wild mice in a volume-dependent fashion, with additional effects of frequency on spleen weight.

Behavioural responses of vertebrates to odours from other species: towards a more comprehensive model of allelochemics

Birte L Nielsen[1], Olivier Rampin[1], Nicolas Meunier[1,2] and Vincent Bombail[1]
[1]INRA UR1197 NeuroBiologie de l'Olfaction, PHASE, Bât 230, 78352 Jouy-en-Josas, France,
[2]Université de Versailles St-Quentin-en-Yvelines, 55 avenue de Paris, 78035 Versailles, France;
birte.nielsen@jouy.inra.fr

The behaviour of an animal can be affected by odours from another species (allelochemics). This phenomenon has traditionally been characterized according to who benefits (emitter, receiver or both) and the odours categorized accordingly (allomones, kairomones, and synomones, respectively), which has its origin in the definition of pheromones, i.e. intraspecific communication via volatile compounds. There exist, however, interspecific odorant effects which do not fit well in this paradigm. Three aspects in particular do not encompass all allelochemical effects: one relates to the innateness of the response, another to the origin of the odour, and the third to the intent of the message. We will present examples of behavioural responses of vertebrates to odours from other species with specific reference to these three aspects. These examples include squirrels anointing themselves with the smell of snakes, and rats responding sexually to female foxes in oestrus. Searching for a more useful classification of allelochemical effects, we examine the relationship between the valence of these odours (attractive through to aversive), and the relative contributions of learned and unconditioned (innate) responses of the affected animals. We propose that these two factors (odour valence and degree of learning) may describe better the nature of interspecific odorant effects than focusing solely on who benefits from the odour based communication.

Long-term stress monitoring of captive chimpanzees in social housings: monitoring hair cortisol level and behaviors

Yumi Yamanashi, Migaku Teramoto, Naruki Morimura, Satoshi Hirata, Miho Inoue-Murayama and Gen'ichi Idani

Kyoto University, Wildlife Research Center, 2-24, Tanaka-sekiden-cho, Sakyo-ku, Kyoto City, 6068203, Japan; yamanashi@wrc.kyoto-u.ac.jp

A vast amount of accumulated evidence suggests that social living is essential to the well-being of captive social animals, including chimpanzees. Although the formation of complex social groups comparable to that of wild groups is recommended, it can sometimes result in increased stress due to factors such as space limitations, male aggression and/or relocation between institutions. In order to maximize the effectiveness of social living, it is important to understand the relationship between such variables and long-term stress levels in captive chimpanzees. We compared behaviors and hair cortisol level (as an index of long-term stress) of captive chimpanzees. Subjects were fifty-nine individuals living in Kumamoto Sanctuary (KS), Kyoto University, comprising:8 relocation group chimpanzees (relocated from another institution as a group), 7 immigrant group chimpanzees (relocated from other institutions and integrated into new social groups within 5 years) and forty-four resident chimpanzees (housed at KS for many years). We cut arm-hair from chimpanzees 3 to 4 times per year. Aggressive and abnormal behaviors were recorded ad-libitum by keepers using a daily behavior monitoring sheet developed for this study. Hair cortisol was extracted and assayed using enzyme imunoassay using a methodology developed in our previous studies. Statistical analyses were conducted using the Generalized Linear Model and model was selected based on likelihood ratio test ($P<0.05$). We found that standardized rate of receiving aggression, rearing history, sex and group formation were found to be significant factors influencing mean hair cortisol level. Sex and group formation were significant factors influencing variance of hair cortisol level. Relocation status was not a significant factor, but mean hair cortisol level was positively correlated with the rate of receiving aggression, especially for males. Additionally, mean and variance of hair cortisol levels in males were higher than those in females (males: 22.9±5.15 pg/mg hair; females: 20.8±3.2 pg/mg hair). These results suggested that the long-term stress level of chimpanzees is related to social factors and males may be more sensitive to such factors than females.

Does semi-group housing of rabbit does remove restrictions on normal behaviour?

Stephanie Buijs[1], Luc Maertens[1], Jürgen Vangeyte[2] and Frank André Maurice Tuyttens[1]
[1]Institute for Agricultural and Fisheries Research (ILVO), Animal Sciences Unit, Scheldeweg 68, Melle, Belgium, [2]Institute for Agricultural and Fisheries Research (ILVO), Technology and Food Science Unit, Van Gansberghelaan 115, Merelbeke, Belgium; stephanie.buijs@ilvo.vlaanderen.be

Reproduction does are commonly housed separately in cages, which have been suggested to restrict intrinsically motivated behaviours (e.g. locomotion and social contact). Using group pens instead of single-doe cages decreases spatial and physical restrictions, but continuous group housing of reproduction does can lead to infertility and high litter mortality. Therefore, we evaluated if semi-group housing (4 does housed together from 18 days after kindling until 3 days prior to the next kindling) would lead to an increase in time spent on behaviours thought to be restricted by single-doe cages. Sixteen does housed in single-doe cages (0.4 m^2 + 0.1 m^2 platform) and 32 does in semi-group pens (2 m^2 + 0.6 m^2 platform/4 does) were observed using continuous sampling during six 30-minute timeslots (immediately after grouping of the semi-group does, 12 hours later, and at midday and midnight 4 and 12 days later). Statistical comparisons were made using Kruskal-Wallis tests and results are presented as medians (interquartile range). In the 30 minutes immediately after grouping, semi-group does spent more time on locomotion than does in single-doe cages (4 (4-5) vs 1 (1-1)%, $P<0.01$), as well as on social sniffing/grooming (1 (1-2) vs 0 (0-0)%, $P<0.01$). Although such behaviours still differed significantly between semi-group and single-doe housing 4 and 12 days after grouping, the differences were much smaller (e.g. midnight 12 days post-grouping locomotion: 0.8 (0.6-1.4) vs 0.2 (0.2-0.3)%, $P<0.05$; social sniffing/grooming: 0.4 (0.4-0.6) vs 0.0 (0.0-0.0)%, $P<0.01$). Attacking/chasing followed a similar pattern (after grouping: 5 (4-8)% in semi-group housing vs 0 (0-0)% in single-doe housing, $P<0.01$; midnight 12 days post-grouping: 0.01 (0.00-0.03) vs 0.00 (0.00-0.00)%, $P<0.10$). Immediately after grouping, semi-group does spent less time in bodily contact with adult conspecifics than does from single-doe housing (who could only lie together with the wire cage wall between them, 2 (1-3) vs 12 (11-15)%, $P<0.01$). This may have been due to the unfamiliarity between the grouped does. However, even 12 days after mixing the time spent in bodily contact was not significantly greater in the semi-group system than in single-doe housing ($P>0.10$). The limited occurrence of locomotion and non-agonistic social contact in semi-group housing suggests that either the does' motivation for such behaviour is limited, or that the specific semi-group system we tested still imposes physical or social restrictions on such behaviour.

Social housing affects the development of feeding behaviour in dairy calves

Emily K. Miller-Cushon[1] and Trevor J. Devries[2]
[1]*University of Florida, Department of Animal Sciences, 2250 Shealy Drive, Gainesville, FL 32601, USA,* [2]*University of Guelph, Department of Animal and Poultry Science, 50 Stone Road East, Guelph, ON, N1G 2W1, Canada; emillerc@ufl.edu*

Social housing for dairy calves influences early feed intake and social interaction, but it is unclear whether it influences post-weaning behaviour. This study investigated how pre-weaning social contact affects feeding behaviour and feeding synchrony after milk weaning. Twenty Holstein bull calves were housed either individually (IH; n=10) or in pairs (PH; n=5) from birth. Calves were offered grain concentrate and milk replacer ad libitum via artificial teats and weaned by incrementally diluting the milk replacer from 39-49 d of age. Post-weaning, IH calves were paired within treatment and all pens were offered a pelleted diet ad libitum and followed until 84 d of age. Feeding times were recorded from video for 3 consecutive days in wk 6, 9, and 12 of age and used to calculate daily meal frequency and meal duration. In wk 9 and 12, frequency and duration of synchronized feeding were also calculated. Data were summarized by week and pen (in wk 6 data were averaged across adjacent pairs of IH calves) and analyzed by time period (pre- and post-weaning) in a general linear mixed model, with week as a repeated measure in the post-weaning period. In addition, preference tests were conducted at time of feed delivery in wk 10 to assess the preference of each calf to feed alongside or out of visual contact of the pen-mate. Prior to weaning, PH calves had more frequent concentrate meals (8.0 vs 4.6 meals/d; SE=0.89; P=0.03) and greater intake (0.17 vs 0.062 kg/d; SE=0.03; P=0.03). Milk intake was not affected by treatment (12.0 l/d; SE=0.87; P=0.8), but milk meals of calves in PH pens were more frequent (7.1 vs 5.0 meals/d; SE=0.55; P=0.02) and smaller (1.85 vs 2.81 l/meal; SE=0.32; P=0.04). After weaning, intake was similar between treatments (6.9 kg/d; SE=0.33; P=0.4), but differences in behavior persisted: calves raised in PH pens continued to have meals that were more frequent (15.3 vs 13.3 meals/d; SE=0.67; P=0.04) and shorter in duration (12.3 vs 15.5 min/meal; SE=0.89; P=0.005). Both treatments had a similar frequency of synchronized meals (7.6 overlaps in feeding/d, SE=0.90; P=0.4). However, when offered a choice to feed alone or alongside their pen-mate, calves raised in PH pens preferred social feeding to feeding alone, whereas calves raised in IH pens had no preference (79.8 vs 58.7% of time on social side during preference test; SE=4.6; P=0.01). These results suggest that meal patterns established in response to different early social environments may persist post-weaning, and that early social contact may have longer-term effects on social feeding behavior.

Freedom to express agonistic behaviour can reduce escalated aggression between pigs

Irene Camerlink[1], Gareth Arnott[2], Marianne Farish[1] and Simon P. Turner[1]
[1]Scotland's Rural College (SRUC), Animal Behaviour & Welfare, Animal Veterinary Sciences Research Group, West Mains Rd., EH9 3JG Edinburgh, United Kingdom, [2]Queen's University, School of Biological Sciences, Institute for Global Food Security, University Road, BT9 7BL Belfast, United Kingdom; irene.camerlink@sruc.ac.uk

Intensive farming limits the opportunities for animals to express their natural behaviour. Natural agonistic behaviour such as threat and withdrawal is important in pigs for establishing and maintaining dominance relationships, but housing is not designed with this requirement in mind. To gain a better understanding of why and when pigs decide to involve in costly fighting we applied a game-theoretical framework using dyadic contests. We studied the importance of non-damaging phases of agonistic behaviour on the time and strategy that pigs take to settle conflicts. Contests (n=52) were staged between unfamiliar pairs of pigs (M/F) of similar age (10 wk) and body weight, in an arena measuring 2.9×3.8 m. Contests lasted until a pig retreated and did not retaliate within 2 min. Behaviour was observed from video and analysed with logistic and linear mixed models. Contests lasted on average 5½ min (339±19 s), from entering the arena until a clear loser was apparent. On average, 87±6 s of the contest was spent on display behaviour, 35±6 s on pushing, and 54±6 s on mutual fighting. Display behaviour was largely showed in a ritualized manner, e.g. parallel walking which is displayed in various species during contests. Pairs that spent more time on display behaviour had a longer contest duration (b=2.4±0.3 s/sec display; $P<0.001$). More display and non-damaging behaviour was seen in contests where an outcome was reached without mutual fighting (15/52 contests). In contests without fights, pigs spent 53% more time in non-damaging investigation of each other ($P=0.06$) and spent 46% more time in parallel walking ($P=0.01$), but 57% less time with their heads up (form of display; $P<0.01$) and 64% less time pushing ($P=0.04$), compared to contests that escalated to fighting. Contests in which no fight occurred showed 2.8 fold more bullying ($P<0.001$). The results suggest that non-damaging ritualized display may facilitate conflict resolution without the need for escalated aggression. Ritualized display does, however, require considerable space to be fully expressed. Commercial space allowance and stocking density may impede this ritualized display and thereby potentially prolong agonistic interactions including bullying behaviour.

Ketoprofen reduces signs of sickness behaviour in pigs with acute respiratory disease

Anna Valros, Outi Hälli, Elina Viitasaari, Minna Haimi-Hakala, Tapio Laurila, Claudio Oliviero, Olli Peltoniemi and Mari Heinonen
Univeristy of Helsinki, Department of Production Animal Medicine, P.O. Box 57, 00014 Helsinki, Finland; anna.valros@helsinki.fi

Respiratory diseases (RD) are common in fattening pigs. In Finland over 10% of pigs were diagnosed with pleurisy or pneumonia at slaughter in 2007. RDs cause both reduced production performance and pig welfare. Earlier studies have shown that pigs with induced viral RD reduce feeding, and increase total resting time as well as proportion of time resting without contact to other pigs. RDs are rarely treated with non-steroidal-anti-imflammatory drugs (NSAIDs), although a quicker recovery would increase welfare as well as production performance. We aimed to evaluate possible effects of orally administered ketoprofen in acute outbreaks of respiratory disease, by performing a double-blinded study on four different farms, all of which had reported acute signs of RD. Actinobacillus pleuropneumoniae was diagnosed in euthanized pigs from all herds. The farm was visited by the research team on day 0, which was within two days of the start of the reported outbreak. Eight pens within a section with RD signs were chosen for observations on each farm. Within each pen five pigs with clear clinical signs of RD were marked to allow for individual direct behavioural observations. The pens were randomly assigned as placebo (P, n=16) or treatment (T, n=16) pens. Pigs were treated with ketoprofen 3 mg/kg (T) or placebo (P) once daily in a small feed ration before morning feeding during d 1-3. The behaviour of the pigs was observed by instantaneous sampling at five min intervals for a total of two hours before afternoon feeding on d 0 and on d 3. Pen means of each behaviour were calculated, and the data analyzed with SPSS 21. Differences in behaviour between d 0 and d 3 was tested with paired t-tests for T and P separately. Effect of treatment on the magnitude of change in behaviour was tested with a univariate model, including farm as fixed factor. The occurrence of several behaviours changed in T pigs from d 0 to d 3: standing, walking, lying sternally, being active, lying alone and exploring increased, while being passive, lying down in total and lying laterally decreased ($P<0.05$ for all). In P pigs the only observed change was a decrease in sternal lying ($P=0.02$). The magnitude of the change in behaviour from d 0 to d 3 was significantly different between P and T groups for standing, lying in total, lying sternal and lateral, and for nosing and passive ($P<0.01$ for all). These results indicate a reduction of behaviour suggested to be indicative of sickness in pigs when treating RD with an orally administered NSAID. Thus, ketoprofen probably enhances recovery and reduces negative impact on pig welfare during acute RD.

Can flunixin in feed alleviate the pain associated with castration and tail docking?

Danila Marini[1,2], Ian Colditz[2], Geoff Hinch[1], Carol Petherick[3] and Caroline Lee[2]
[1]Environmental and Rural Science, University of New England, 2350, Armidale, NSW, Australia,
[2]CSIRO, FD McMaster Laboratory, New England Highway, 2350, Armidale, NSW, Australia,
[3]The University of Queensland, Brisbane, St Lucia, 4072, QLD, Australia; danila.marini@csiro.au

Australian lambs undergo the routine invasive husbandry procedures of castration and tail-docking (marking) as part of standard farming practice and currently do not receive pain relief for these procedures due to lack of registered analgesics. This research tested the efficacy of an analgesic (flunixin) administered in pelleted feed following marking of lambs. Thirty-two, single Merino lambs, aged 6-10 weeks were randomly assigned to 4 treatments: Sham (S); marking, no pain relief (C); marking, flunixin in feed (F); and marking, flunixin injected (I). Group F received flunixin (4.0 mg/kg) at 24 h and 1 h prior to marking. Group I received flunixin (2.0 mg/kg) 1 h prior to marking. Lambs were filmed post-marking; behavioural indicators of acute pain, which included restlessness, foot stamping, rolling, jumping and easing quarters, were recorded for each lamb for 1 min at 5-min intervals for the first h post-marking. Blood samples for haematology and cortisol assay were taken at 0 h and 30 min, 6, 12, 24 and 48 h post-marking. Inflammation was measured using neutrophil counts. Data were analysed using linear and non-linear mixed effects model, with behavioural and cortisol data log transformed. Treatment had a significant effect on the total number of pain-related behaviours ($P<0.001$); S group showing significantly less than lambs in the other groups (Log mean ± SEM, 0.329±0.066, back transformed mean, 1.135). F and I lambs exhibited less pain-related behaviours (0.556±0.066, 2.597 and 0.557±0.066, 2.608) compared with C lambs (0.810±0.066, 5.451). C, F and I lambs showed a significant increase compared to time 0 in cortisol at 30 min after marking ($P<0.001$), it was also significantly higher compared with S lambs ($P<0.001$ for all groups). At 6 h group I had lower cortisol concentrations than group C and at 48 h group F had lower concentrations than group C. The administration of flunixin to group F and I significantly reduced inflammation ($P<0.05$) following marking at 6, 12 and 24 h compared to C lambs. Group F had lower neutrophil counts than group I at 12 and 24 h ($P<0.05$). The results indicate that administering flunixin to lambs orally in feed is as effective as injecting flunixin, indicated by reduced inflammation and pain responses in lambs following marking. Provision of flunixin in the feed provides a practical way to provide pain relief to lambs following painful procedures by removing the need for multiple injections and reducing handling stress and labour requirements.

The effect of ketoprofen administered post farrowing on the behaviour and physiology of gilts and sows

Sarah H. Ison and Kenneth M. D. Rutherford

SRUC, Scotland's Rural College, Animal Behaviour & Welfare, Roslin Institute Building, Easter Bush, Roslin, EH25 9RG, Midlothian, United Kingdom; sarah.ison@sruc.ac.uk

Recent research into the use of non-steroidal anti-inflammatory drugs (NSAIDs) post farrowing demonstrates some benefits to sow health, welfare and piglet performance. These studies focused on general measures of welfare and indicated that some individuals may benefit more than others. This study investigated the use of the NSAID ketoprofen in a randomised, blinded, placebo controlled trial using 24 gilts and 32 sows housed in conventional farrowing crates. Individuals were randomly allocated to receive ketoprofen (3 mg/kg bodyweight) or the equivalent volume of saline by intramuscular injection 1.5 hours after the birth of the last piglet. Saliva was sampled for cortisol following the morning and afternoon feed for two days before farrowing, the day of farrowing and days one, two, three, five and seven post-farrowing. An additional sample was taken 6 hours following the injection of ketoprofen or saline. Potential pain-related behaviours associated with farrowing, identified in a previous study, were recorded. These included tremble (shivering as if cold), back leg forward (holding the back leg forward and/or in towards the body) and back arch (legs tensed and pushed away from the body forming an arch in the back). Behavioural observations were made for 15 minutes every 1.5 hours from one hour after the birth of the last piglet for eight observations. For the 24 gilts, a further three 15 minute observations were made every three hours following the previous eight. Data were analysed using linear mixed models in Genstat. There was no difference in cortisol by treatment ($P>0.05$) or treatment × time collected ($P>0.05$). However, salivary cortisol tended to differ overall between gilts and sows (Gilts=3.45±0.16, Sows=3.63±0.12: $P=0.07$) and there was a significant gilt/sow × time collected interaction ($P=0.012$) as sows had higher salivary cortisol on the day of farrowing and the immediate post-farrowing period. None of the behavioural variables had an overall treatment difference. However, for the gilts, a tendency for a treatment × observation number interaction ($P=0.06$) was present for tremble. More trembling was observed in the control group for four of the 15 minute observations, which were between 8.5 and 16 hours after the injection was administered. The lower salivary cortisol level seen by gilts compared with sows on the day of farrowing and the immediate post-farrowing period indicates lower physiological stress, which could be due to a smaller and lighter litter size. Less trembling seen in gilts treated with ketoprofen for several of the post-farrowing observations could be indicating an improvement in recovery post-farrowing.

Is social rank associated with health of transition dairy cows?

Marcia Endres, Karen Lobeck-Luchterhand, Paula Basso Silva and Ricardo Chebel
University of Minnesota, 1364 Eckles Avenue, St. Paul, MN 55108, USA; miendres@umn.edu

The objective of this study was to determine whether social ranking was associated with health of transition dairy cows. A total of 953 Jersey cows were used in the study. They were housed in a sand-bedded freestall barn with maximum stocking density of 109%. Cows were examined on days in milk 1, 4, 7, 10, and 13 for the diagnosis of retained fetal membranes (RP) and metritis. Metritis was defined as cows with watery, pink or brown, and fetid uterine discharge; acute metritis included cows that had a fever (> 39.5 °C). Cows were classified with subclinical ketosis when BHBA concentration was $\geq$1,200 µmol/l. Cows were observed once daily for displacement of abomasum (DA) and three times daily for mastitis. Displacements from the feed bunk were measured continuously for 3 hours after fresh feed delivery 4 days/week during the 4 weeks prior to calving for determination of social rank. A displacement index (DI) was calculated for each cow as the number of displacements as actor (cow initiated the displacement) divided by total displacements as actor or reactor (cow received the displacement). Cows were categorized into social rankings: cows with a DI<0.4 were considered low-ranking, 0.4 to 0.6 were considered middle-ranking, and >0.6 were considered high-ranking. Proc Logistics in SAS was used to investigate the association between social ranking and health. There was no association of dominance category with incidence of metritis, subclinical ketosis, displaced abomasum, endometritis, acute metritis, or mastitis. Retained fetal membrane was the only peripartum health event associated with dominance category (P=0.041). High-ranking cows were 2.0 times more likely to have RP than low-ranking cows with no differences between low- and middle-ranking cows. A cow's ability to displace another cow from the feed bunk during the prepartum period as a determinant of her social rank was not very helpful in predicting the odds of having a transition health disorder.

Risk factors associated with stranger-directed aggression in dogs

Hannah Flint[1], Jason Coe[1], James Serpell[2], David Pearl[1] and Lee Niel[1]
[1]*University of Guelph, Population Medicine, 50 Stone Rd E, Guelph ON, N1G2W1, Canada,*
[2]*University of Pennsylvania, Clinical Studies, 3800 Spruce St, Philadelphia, PA 19104, USA;*
flinth@uoguelph.ca

Aggression in canines is a safety concern both for humans and animals, and can lead to decreased animal welfare in affected dogs due to abuse, neglect, relinquishment or euthanasia. Our objective was to explore risk-factors for stranger-directed aggression in dogs using a previously validated, owner-completed canine behaviour questionnaire (C-BARQ). Scores for stranger-directed aggression, stranger-directed fear, non-social fear and touch sensitivity were calculated using factors developed by Hsu & Serpell and were then dichotomized. Data were analyzed using multivariable logistic regression models, with household as a random effect (n=15,911 dogs). Dogs were more likely to be aggressive if they were male (OR=1.32; $P<0.001$), neutered/spayed (OR=1.38; $P<0.001$), touch sensitive (OR=1.29; $P<0.001$) or fearful of strangers (OR=5.07; $P<0.001$). There was also an association with breed group ($P<0.001$), where the dog was acquired ($P<0.001$) and the dog's role ($P<0.001$). Dogs were less likely to be aggressive with increased age at acquisition (OR=0.864/yr; $P<0.001$). The random effect for household was significant ($P<0.001$) indicating that there was some correlation in behaviour among dogs within the same household; suggesting household effects may need to be examined further. When looking only at dogs categorized as aggressive towards strangers (n=12,369), dogs were more likely to be categorized as having severe aggression (biting or attempting to bite) if they were male (OR=1.45; $P=0.019$), a toy breed (vs mixed/other; OR=1.92; $P=0.006$), or had non-social (OR=1.85; $P<0.001$) or stranger-directed fear (OR=2.46; $P<0.001$). There was also an association with the dog's role ($P<0.001$). In this model, there was no significant effect of household. These results suggest that while variables related to environment and experience play a role in whether dogs are aggressive, the factors associated with whether an aggressive dog attempts to bite are primarily related to inherent traits of the dog itself.

Assessing stress and behaviour reactions of dogs at the veterinary clinic

Ann-Kristina Lind[1,2], Eva Hydbring-Sandberg[3], Björn Forkman[4] and Linda J. Keeling[1]
[1]*Swedish University of Agricultural Sciences, Department of Animal Environment and Health, P.O. Box 7068, 750 07, Uppsala, Sweden,* [2]*Swedish Institute of Agricultural and Environmental Engineering, P.O. Box 7033, 750 07, Uppsala, Sweden,* [3]*Swedish University of Agricultural Sciences, Department of Anatomy, Physiology and Biochemistry, P.O. Box 7011, 750 07, Uppsala, Sweden,* [4]*University of Copenhagen, Department of Large Animal Sciences, Faculty of Health and Medical Sciences, Grønnegårdsvej 8, 1870, Frederiksberg, Denmark; ann-kristina.lind@jti.se*

The aim of this study was to develop a simple, non-invasive scoring system to assess stress in dogs at the clinic by the use of questionnaires and behaviour tests. Questionnaires were filled in by the owner, test-leader, nurse and veterinarian. To assess the dog-owner relationship, the Monash Dog Owner Relationship Scale was filled in by the dog-owner. The inter-observer agreement between the observers was calculated with prevalence-adjusted bias-adjusted kappa (PABAK) and the calculated agreements were good to excellent (PABAK>0.61). Three behaviour tests were carried out to describe the dog's reaction in the veterinary clinic; a 'social contact' test, a 'play' test and a 'treat' test. The play and treat tests were carried out both inside and outside the veterinary clinic to see if the dogs reacted differently in the two situations. Comparisons were made using one-way ANOVA tests. There were 233 dog-owners who took part in the questionnaire study and 105 of their dogs (patients at the clinic) were tested. The dog-owner, the test-leader, the nurse and the veterinarian were each asked to score if the dog was stressed and to evaluate overall how the dog experienced the visit. Dogs rated as more stressed were less likely to take social contact with an unfamiliar person ($P<0.0001$) and more willing to play and eat a treat outside the veterinary clinic compared to inside the clinic ($P<0.0001$). The results also indicated that the type of relationship the owner has with the dog may influence the dog's behaviour during the clinical examination. Dog-owners who had a high perceived emotional closeness with their dog or a high perceived awareness of the cost of having a dog were scored by the veterinarian as tending to be worse at calming their dog during the clinical examination. Finally the good agreement between the different measures in this study suggests that there is potential for a scoring system to be developed to assess the extent to which the dog is stressed in the clinic. Awareness of dog's different reactions to a veterinary clinic is important for them to be handled accordingly (for their own safety and for that of the examining veterinarian), and so that their overall experience of the clinic is improved.

Association of the behaviour of Swedish companion cats with husbandry and reported owner behaviours

Elin N Hirsch[1], Johanna Geijer[2], Jenny Loberg[2], Christina Lindqvist[2], James A Serpell[3] and Maria Andersson[2]

[1]Swedish University of Agricultural Sciences, Animal Environment and Health, Växtskyddsvägen 3, P.O. Box 103, 230 53 Alnarp, Sweden, [2]Swedish University of Agricultural Sciences, Animal Environment and Health, P.O. Box 234, 532 23 Skara, Sweden, [3]University of Pennsylvania, School of Veterinary Medicine, 3900 Delancey Street, Philadelphia, PA 19104-6010, USA; elin.hirsch@slu.se

Occurrence of undesired behaviours (UB) has previously been shown to increase the risk of relinquishment or abandonment of a companion animal, and UB has been shown to be connected to owner knowledge and certain husbandry practises. To investigate the relationship between cats and their owners in Sweden, a net based questionnaire was distributed. The survey was a Swedish translation of the Fe-BARQ (© James A. Serpell, University of Pennsylvania) containing additional questions regarding the cat-human relationship. The main aim with this part of the survey was to see if owner perception of cats relate to occurrence of undesired behaviours such as elimination problems. The secondary aim was to see if the time a cat was left alone (TLA) influenced occurrence of reported UB. Questions regarding owner perception of cats were posed using a Likert-like 3 point scale and TLA questions on a numerical scale. The survey was open between 29 April and 1 July 2014, and of 4,209 responses 78% (3,280) were completed. Preliminary results showed that 82% of owners reported no perceived behavioural or temperamental problems (BTP). Of the 3280 described cats, 50% were domestic short hair, 53% were males, 90% were neutered and 33% had free outdoor access. It was most common to leave the cat alone for 6-8 h/day (25%) followed by 4-6 (20%). A chi-square test of independence was used to investigate the association between outdoor access and owner perceived BTP. There was an association between outdoor access and perceived problems χ^2 (1, n=2,935)=11.348, P=0.001 in that cats with outdoor access had lower observed reporting of BTP than expected. There was an association between having studied ethology at university level and if you agreed to the statement that cats misbehave out of spite χ^2 (1, n=3,279)=38.381, P=0.000, showing that having studied ethology resulted in fewer observed agreements that cats act out of spite than expected by the chi-square. No association was found between TLA and BTP χ^2 (2, n=3,280)=1.618, P=0.445. As with previous studies from other countries we found an association between knowledge and outdoor access and perceived BTP. We did not find a relationship with TLA. One weakness with the present data could be that only dedicated owners took the time to participate, however this is the first large demographic study of house cats in Sweden.

Pets in the digital era: live, robot or virtual

Jean-Loup Rault
Animal Welfare Science Centre, University of Melbourne, Parkville 3010, Australia; raultj@unimelb.edu.au

Are we still going to care for animals in the digital age? Over half the people in Western societies share their daily life with pets, hence the norm rather than the exception. Our shared history with domestic animals goes back tens of thousands years. However technological advances in the last decades -computer, internet, social media- revolutionised our means of communication, and particularly our social lives. These technologies are likely to change human-animal relationships, and concurrently the place of animals in human societies. This conundrum could be addressed according to the concept of the 3Rs: (1) refinement – not all species are suitable or animal use ethical; (2) reduction – pet ownership as luxury, remote interactions with animals through technology; and (3) replacement – robot, virtual reality. Refinement has been suggested in that highly sentient or intelligent species are not suitable for captivity as we are unable to fulfill their social or mental needs. Technologies could however improve animal welfare, by providing environmental enrichment or facilitate interactions such as remote communication. Reduction is an interesting proposal, as there is an inherent conflict between our remoteness from nature which seems to stimulate pet keeping, and sustainability of pet keeping in a growing, urbanised society. Alternatives that have been investigated are remote internet interactions with farm animals for example. The last option is most intriguing, that technologies could allow us to replace animal use. There are already alternatives to the use of animals in research with in-vitro or computer models, synthetic fibres, and in its infancy for meat production from cells. Whether technologies could replace live pets however remains mostly unexplored. Pet robots appear to elicit similar responses from humans as live pets, but behaviours that mimic a live interaction need to be refined. Progress is on-going to computer-simulate interactions with a pet through virtual reality. The Tamagotchi, Sony AIBO robotic dogs or Nintendogs (Japanese-led field!) still represent mediocre substitutes to live pets. However, quick technological progress is to be expected, and should partly rely on ethology to answer some of the practical and fundamental questions raised along the way. What is the behaviour of animals (including humans) toward technology? What behaviours and cues are necessary for an interaction, both from an emotional and communicative point of view? Is applied ethology missing on the largest revolution in human-animal relationship history, its development and consequences? Ethologists need to be part of this inter-disciplinary approach to address the future of human-animal relationships, and in the long-term of animal welfare science.

Playful handling before blood sampling improves laboratory rat affective state

Sylvie Cloutier[1] and Ruth Newberry[2]
[1]Canadian Council on Animal Care, 800-190 O'Connor Street, K2P 2R3,Ottawa, ON, Canada, [2]Norwegian University of Life Sciences, Department of Animal and Aquacultural Sciences, P.O. Box 5003, 1432, Ås, Norway; scloutier@ccac.ca

We hypothesized that experiencing playful handling (tickling) prior to blood collection increases positive, and reduces negative, affective states in laboratory rats. Female and male pair-housed Long-Evans rats (n=112), 30-60 days of age, were habituated to tickling (HT), or not handled (HN), for 2 min daily for 3 days prior to sampling. They were assigned to one of two blood collection treatments administered once in each of two consecutive weeks: (1) tickling immediately followed by blood collection (TB); (2) no handling followed by blood collection (NB). Rat responses on collection days were investigated by measuring production of 50- and 22-kHz ultrasonic vocalizations (USVs) (indicators of positive and negative affective states, respectively), and audible vocalizations (indicating discomfort), during the procedure. Production of 50-kHz USVs was marginally affected by the interaction between habituation treatment and blood collection treatment ($P=0.058$). Rats habituated to tickling and tickled immediately before blood collection (HT-TB) tended to produce more 50-kHz USVs than rats in the other conditions (mean±SE, HN-NB: 4.0±0.82 calls/min, HN-TB: 15.9±2.58, HT-NB: 5.5±1.03, HT-TB: 12.0±1.84). Habituation to tickling, blood collection treatment and collection day affected rate of audible calls. Habituation to tickling increased audible call production compared to no handling (HN: 1.4±0.26, HT: 2.2±0.29, $P=0.02$), while tickling immediately before blood collection reduced audible calling (NB: 2.0±0.27, TB: 1.5±0.29, $P=0.005$). Rats called more on collection day 1 than 2 (day 1: 2.2±0.33, day 2: 1.4±0.20, $P=0.01$). Production of 22-kHz USVs was rare (0.1±0.03 calls/min) and unaffected by treatments, and no sex differences were detected. Our results suggest that tickling immediately before blood collection contribute to inducing a more positive affective state in rats during blood sampling.

The effect of visitor number on kangaroo behaviour and welfare in free-range exhibits

Sally Sherwen[1,2], Paul Hemsworth[1], Kym Butler[1,3], Kerry Fanson[4] and Michael Magrath[2]
[1]*University of Melbourne, Animal Welfare Science Centre, Parkville, Victoria 3010, Australia,* [2]*Zoos Victoria, Wildlife Conservation and Science, Melbourne Zoo, Parkville, Victoria 3052, Australia,* [3]*Department of Economic Development Jobs Transport and Resources, Hamilton, Victoria, 3300, Australia,* [4]*Deakin University, Centre for Integrative Ecology, Waurn Ponds, Victoria 3216, Australia; ssherwen@zoo.org.au*

Free range exhibits are becoming increasingly popular in zoos. These exhibits typically have no physical barrier separating animals and visitors and very little research has been conducted on the impacts of visitors on animal behaviour and stress in these exhibits. We investigated the effects of visitor number on the behaviour and stress physiology of Kangaroo Island (KI) Kangaroos, *Macropus fuliginosus fuliginosus*, and Red Kangaroos, *Macropus rufus*, housed in two free range exhibits in Australian zoos. Kangaroo behaviour and faecal glucocorticoid metabolite (FGM) concentration were studied over 12 days at Melbourne Zoo and 14 days at Healesville Sanctuary. Behavioural observations were conducted on individual kangaroos at each site using instantaneous point sampling to record activity (e.g. vigilance, foraging, resting) and distance from the visitor pathway. Individually identifiable faecal samples were collected at the end of each study day and analysed for FGM concentration. Regression analyses were used to examine relationships between visitor number and kangaroo variables. To ensure residuals were reasonably independent, data were averaged for each set of consecutive days (with each set of days at least 5 days apart). These blocks of days were the unit of analysis. Kangaroo Island Kangaroos increased the time spent engaged in visitor-directed vigilance ($F_{1,8}=13.88$, $P=0.006$) and locomotion ($F_{1,8}=9.34$, $P=0.02$) and decreased the time spent resting ($F_{1,8}=11.58$, $P=0.009$) when visitor number increased. Red Kangaroos also increased the time spent engaged in visitor-directed vigilance ($F_{1,5}=17.38$, $P=0.009$) when visitor number increased. There was no effect of visitor number on the distance kangaroos positioned themselves from the visitor pathway (KI: $F_{1,8}=0.19$, $P=0.67$, Red: $F_{1,5}=3.96$, $P=0.10$) or FGM concentration (KI: $F_{1,6}=0.80$, $P=0.40$, Red: $F_{1,3}=2.45$, $P=0.22$) in either species. Therefore, while there was some evidence of changes in kangaroo behaviour, such as increased vigilance and reduced resting in response to increasing visitor numbers, there was no evidence of adverse effects on animal welfare based on avoidance behaviour or stress physiology in these study groups.

Can visible eye white be used as an indicator of positive emotional state in cows?

Helen Proctor and Gemma Carder
World Animal Protection, 5th Floor, 222 Gray's Inn Rd, London, WC1X8HB, United Kingdom;
helenproctor@worldanimalprotection.org

Reliable measures of positive emotions in animals are widely needed. We explored whether the percentage of visible eye whites in dairy cows is a valid measure of a low arousal positive emotional state, by using stroking as the positive stimulus. We performed 372 full 15 minute focal observations on 13 habituated cows, filming the focal cow's eye and recording their behaviours during three phases of the focal observation: pre-stroking, stroking and post-stroking. We then calculated the percentage of visible eye white at nine pre-determined measurement points; three in each phase of the focal observation. We analysed the visible eye white data using the One-Way Repeated Measures ANOVA test. We found that the percentage of visible eye white dropped significantly during stroking, compared with during both the pre-stroking and post-stroking phases ($F_{(1.242, 14.9)}$=4.32, P=0.025). We controlled for high arousal behaviours, and we analysed the performance of behaviours known to be associated with positive emotions in cattle. 'Stretching neck' for example, was performed significantly more during stroking than during pre-stroking or post-stroking (X^2=700.68, P=0.000). This and the performance of affiliative behaviours such as 'leaning into stroker' provide support for the use of stroking as a stimulus to induce a positive emotional state in this study. This study has built upon existing work in this field to provide greater insight into the use of visible eye whites as a measure of emotional state in cattle. Our study specifically focused on the effect of a change in valence, rather than on a change in arousal as done previously. Our results therefore provide helpful insight into the relationship between visible eye whites, valence and arousal and support previous studies which suggest that visible eye whites may serve as a dynamic measure of emotional state in cows.

Nasal temperatures and emotions; is there a connection in dairy cows?

Helen Proctor and Gemma Carder
World Animal Protection, 5th Floor, 222 Grays Inn Rd, WC1X 8hb, United Kingdom;
gemmacarder@worldanimalprotection.org

Previous studies have indicated that peripheral temperatures in animals are connected with high arousal, negative experiences, with the states; fear, stress and frustration, resulting in a drop in peripheral temperature. In this study we explored whether nasal temperatures in dairy cows were affected by their emotional state. We conducted experimental stroking on 13 habituated cows to induce a low arousal, positive emotional state. Each of the 350 full focal observations comprised of three 5 minute segments; pre-stroking, stroking, and post-stroking, during which time the focal cow's nasal temperature was recorded using an infra-red thermometer gun at six points, twice during each segment. The One-Way ANOVA repeated measures test found a significant difference overall between the three segments ($F_{(2, 1.94)}$=9.37, $P<0.01$), and the post-hoc pairwise comparisons found a significant decrease in the total mean nasal temperature during the stroking segment (25.91 °C, SD=1.21), compared with both the pre-stroking (26.27 °C, SD=1.01, $P<0.01$) and post-stroking segments (26.44 °C, SD=1.12, $P<0.01$). We found no significant difference in temperature between the pre-stroking and post-stroking segments ($P=0.14$). The cows were considered to be in a low state of arousal throughout the focal observation period as no changes to their environment were made, high arousal behaviours were controlled for, and the cows were fully habituated. Furthermore, previous research has shown stroking to be a positive experience for cows, and to lower both their heart rate and cortisol levels. We suggest therefore that the drop in nasal temperature found during stroking was a result of a change in emotional valence, rather than a change in arousal. This leads us to conclude that a positive emotional state may also cause the peripheral temperature to drop in mammals in the same way a negative state does. Nasal temperatures may therefore serve to be a useful tool in measuring a change in emotional state in cattle.

The body talks: body posture as indicator of emotional states in dairy cattle

Daiana De Oliveira, Therese Rehn, Yezica Norling and Linda J. Keeling
Swedish University of Agricultural Sciences, Dept. of Animal Environment and Health, Box 7068,
Uppsala, Sweden; daiana.oliveira@slu.se

We aimed to evaluate cow's body posture and heart rate (HR) as an indicator of their emotional state. Our 4 treatments were based on 2 different biological systems; feeding and tactile stimulation. During Feed given (FG) cows were given a portion of concentrate and during Feed missing (FM) they saw their neighbour receive feed, but missed to receive anything themselves. During Tactile brush (TB) cows were brushed by a handler following a specific procedure and during Tactile water (TW) they were squirted with a small jet of water at the same locations and rate as they were brushed. We used a balanced repeated design, in which all cows (12 Swedish Red in individual tie stalls) experienced all 4 treatments during a 4-week period. Treatments were given in 2 sessions/day for 7 min on each occasion (1 min pre-treatment (a) + 5 min treatment (b) + 1 min post-treatment(c)). From videos, tail and stepping movements were recorded continuously and instantaneous observations at 5 s intervals were used to record ear, head and neck positions and other behaviours. Data were analyzed using GLMM models in PROC GLIMMIX. Significant interactions were found between treatments and body positions for head, ears and neck ($P<0.001$, respectively). TB and TW elicited similar body postures; head turned more to the right, ears in the back up position, neck below horizontal and a higher frequency of rumination ($P=0.01$). During TB cows also had their head more often in the neck down position, suggesting a relaxed state as shown previously. During the FG treatment cows faced forwards and had their neck down (because they were eating from the feed trough), but they also had their ears most often in the axial position, showed more head tilting ($P<0.001$) and stepping with their back legs ($P<0.001$). FM did not elicit any unique position of ear, head and neck; however there was increased oral behaviour ($P<0.001$) and stepping of the front legs ($P<0.001$). HR was highest in the FG treatment during periods b and c (a=81.9±0.7; b=87.6±0.3; c=88.6±0.7, $P<0.001$, mean±SE). Our results support what has been found before in free moving cows, regarding relaxed ears indicating positive states, but also that other parts of the body are important when assessing emotional states. These results suggest that body posture could be a potential on-farm welfare assessment measure in cattle and they help provide a basis for future research in the field of positive welfare indicators.

Effects of intranasal administration of arginine vasopressin in Holstein steers

Masumi Yoshida[1], Kosuke Momita[2], Masayoshi Kuwahara[1], Etsuko Kasuya[3], Madoka Sutoh[4] and Ken-ichi Yayou[3]
[1]*The University of Tokyo, Bunkyo-ku, Tokyo, 113-8657, Japan,* [2]*Tokai University, Aso-gun, Kumamoto, 869-1404, Japan,* [3]*National Institute of Agrobiological Sciences, Tsukuba, Ibaraki, 305-8602, Japan,* [4]*NARO Institute of Livestock and Grassland Science, Tsukuba, Ibaraki, 305-0901, Japan; masumiyoshida@affrc.go.jp*

Livestock stress reduction is important for animal welfare. Intranasal oxytocin suppresses stress responses or produces a sedative effect. We previously suggested that oxytocin intranasal administration has a sedative effect in Holstein steers. Reports speculated that the central effects of oxytocin are induced via activation of arginine vasopressin (AVP) V1a receptors expressed in olfactory nerves. We therefore investigated the effects of intranasal AVP administration in Holstein steers. Five Holstein steers (7-9 months) were used. At experimental point 0, 1 ml saline (SAL), AVP 100 nmol (AVP100) or AVP 200 nmol (AVP200) /1 ml saline was sprayed into both nostrils of each calf. Behaviour was continually sampled for 90 min and was categorized as standing without ruminating (S), standing with ruminating (SR), lying without ruminating (L), and lying with ruminating (LR). The S, SR, L, and LR time ratios, and the total frequencies of maintenance and abnormal behaviours were compared among treatments. Electrocardiograms were continuously recorded. We evaluated autonomic nervous activities using heart rate variability obtained by power spectrum analysis. Low frequency (LF; 0.04-0.1 Hz) and high frequency (HF; 0.1-0.8 Hz) powers were calculated. The L time ratio was significantly less in AVP200 (34.8±9.0%; Scheffe test, $P<0.05$) than in SAL (45.3±7.8%). The LR time ratio was significantly greater in AVP200 (46.8±9.3%) than in SAL (23.4±9.2; $P<0.05$). The number of head shaking, abnormal behavior, was significantly less in AVP200 (1.0±1.7) than in SAL (4.4±3.2; Scheffe test, $P<0.05$). There was a significant interaction between time and treatments in HF (two-way ANOVA, $P<0.05$), which was significantly higher in AVP200 at 70 and 80 min after administration than in SAL (Fisher's least significant difference test, $P<0.001$). Intranasal AVP increased rumination. Reduced rumination has been reported in some stressful situations. Any potential stress caused by the rearing environment may have been reduced by intranasal AVP. Decreased abnormal behaviour and the increase in parasympathetic tone (increased HF) in the present study also could indicate a potential stress reduction effect of intranasal AVP. These findings suggest that intranasal AVP had an anti-stress effect on the steers and that it could become a useful method for reducing cattle stress.

Effect of intranasal oxytocin administration on serum oxytocin and its relationship with responses to a novel material on cow

Siyu Chen, Sanggun Roh and Shusuke Sato
Graduate School of Agricultural Science, Tohoku University, 232-3, Yomogita, Naruko-Onsen, Osaki, Miyagi, 989-6711, Japan; chensiyu140@gmail.com

In our previous study, we identified that serum oxytocin (OT) concentrations positively correlated with exploratory behavior. Further, it was reported that nasal OT administration had an effect on social behavior and stress in many humans' studies. The aim of the present study is to investigate whether nasal OT administration has an effect on serum OT concentration, social behavior and responses to a novel material in cows. 15 Japanese Black cows were used. They were randomly assigned to 200 µg/l OT (200 OT), 100 µg/l OT (100 OT) or saline (NS) intranasal administration groups. (1) On the test day, each cow was fitted with an indwelling jugular catheter. After they calmed down, 0.5 ml of OT or saline solution was sprayed into each nostril. Then blood samples were collected at -5, 3 and 10 min, and analyzed for OT and cortisol concentrations by using ELISA. Time 0 was considered as the time when the solution was completely administered. (2) On the novel material test day, immediately after the administration of OT or saline, cow was released into the experimental pen with a novel decoy of a Holstein calf. Numbers of self-grooming, approaching to the decoy, sniffing the decoy and investigative behavior to the environment, duration of self-grooming and sniffing the decoy in Sec were recorded for 20 min. OT and cortisol concentrations were analyzed by two-ways repeated measures ANOVA. Correlation coefficients were calculated among the serum OT, cortisol concentration and behaviors. (1) 200 OT increased serum OT concentration (200 OT: 32.7±13.1, 54.8±9.4, 42.9±6.1; 100 OT: 47.5±19.9, 49.9±8.8, 61.5±12.6; NS: 36.6±7.5, 40.2±13.6, 41.1±7.2 at -5, 3 and 10 min, Mean ± SD, pg/ml). The interaction between the treatment and elapsed time tended to be different (F=2.27, P=0.09). Serum OT concentration was significantly higher at 3 than -5 and 10 min in 200 OT (F=4.27, P=0.03). No correlation was found between serum OT and cortisol concentration. (2) All the behaviors were not different among the three treatments. The number of approaching to the decoy was significantly correlated with the serum OT concentration of -5 min (R=0.76, P=0.01). Duration of self-grooming and the number of sniffing the decoy tended to correlate with the serum OT concentration of -5 min, respectively (R=0.56, P=0.06; R=0.52, P=0.08). Nasal OT administration increased serum OT concentrations, though it didn't influence social behaviors and response to the novel material. We did confirm that these behaviors are controlled by the basal serum OT concentration.

Inhaled oxytocin promotes social play in dogs

Teresa Romero[1], Miho Nagasawa[2], Kazutaka Mogi[2], Toshikazu Hasegawa[1] and Takefumi Kikusui[2]
[1]*The University of Tokyo, Department of Cognitive and Behavioral Sciences, 3-8-1 Komaba, Meguro-ku, 153-8902 Tokyo, Japan,* [2]*Azabu University, Department of Animal Science and Biotechnology, 1-17-71 Fuchinobe, Chuoku, Sagamihara, 252-5201 Kanagawa, Japan; tromero@darwin.c.u-tokyo.ac.jp*

Play is a highly plastic and versatile behavior that generally occurs when animals are free from environmental and social stressors. Recent evidence shows that through play animals reduce tension around stressful situations, or turn a stranger into a familiar individual. Furthermore, play is frequently used as part of therapies to correct behavioral problems in dogs. Thus, the combination of behavioral interventions with pharmacotherapies that promote affiliation in general and play in particular might be a potential fruitful strategy for the treatment of behavior problems in companion animals. In a recent paper, we examined whether oxytocin in the domestic dog modulates the maintenance of close social bonds in non-reproductive contexts. We found that exogenous oxytocin promotes positive social behaviors not only towards conspecifics, but also towards human partners. In the present study we examined in further detail the effect that oxytocin manipulation has on social play. A total of sixteen adult dogs from different breeds participated as subjects in a randomized placebo-controlled experiment (females=8; male=8; mean age 6.1 yr. (SEM=0.7)). Each dog received a nasal spray of 100 µl of oxytocin (40 IU, Peptide institute, Japan) or 100 µl of saline solution, depending on the testing condition. All subjects received both conditions and each condition was carried out on different days. The inhalation of this dose of oxytocin in dogs has no side effects. After spray intake, dogs stay in the experimental room (11.5×6.5 m) with their owners and a familiar dog partner, and their behaviors were video recorded for 60 minutes. When sprayed with oxytocin, subjects initiated play sessions more often (Wilcoxon signed rank test z=-1.997, n=15, P=0.046) and played for longer periods of time (z=-2.040, n=15, P=0.041) than when sprayed with saline. Furthermore, after oxytocin nasal intake dogs displayed play signals more often than after saline administration (z=-2.090, n=15, P=0.037), suggesting that oxytocin enhances dogs' play motivation. To our knowledge, this study provides the first evidence that oxytocin promotes social play in the domestic dog. We use these results to hypothesize on the potential therapeutic use of oxytocin for promoting social behaviors and treating social deficits in the domestic dog.

Further examination of behavioural test for kittens: comparison of impression between observers and across situations

Nodoka Onodera, Yoshihisa Mori and Yoshie Kakuma
Teikyo University of Science, Department of Animal Sciences, 2-2-1 Senjusakuragi, Adachi-ku, Tokyo, 120-0045, Japan; onodera@ntu.ac.jp

There is an increased demand for promoting adoption of relinquished cats, especially kittens, at animal shelters. Although it is important for the shelter to pass information on the kitten's temperament to new owners for successful adoption, few standardized behavioural tests have been established and used for kittens. In this study, we carried out the behavioural test we previously developed with more kittens for further examination. In addition, we compared impression of kittens' personality evaluated by caring staff and the experimenter, and in a cage during care time and in a novel place. In total the behavioural test was carried out for eighty-eight kittens (42 males and 46 females) between 1-5 months of age before adoption. The test examined the responses of kittens to novel place, toys, and unfamiliar person, and the number and latency for each response were recorded. Among these kittens, twenty-four (13 males and 11 females) were evaluated by both caring staff and the experimenter in the cage during care time as impression with a scale of 1 to 7 in terms of fearfulness, activeness, aggression and affection-demanding. The impression of the same kittens were evaluated again by the experimenter while the behavioural test was performed in a novel room. The principal component analysis extracted three elements in kittens' responses at the test with the cumulative contribution ratio as 69.6%. The components were labeled as 'Alert,' 'Bold,' and 'Reactive.' Significant positive correlations were found between the staff's and the observer's impression in 'fearfulness,' (rs=0.847, $P<0.01$) 'aggression' (rs=0.649, $P<0.01$) and 'affection-demanding' (rs=0.519, $P<0.05$), but not in 'activeness.' Significant differences were found between their impressions in 'activeness,' 'aggression' and 'affection-demanding' by Wilcoxon signed-rank test ($P<0.05$). Significant positive correlations were found between the observer's evaluation of 'fearfulness' (rs=0.855, $P<0.01$) and 'activeness' (rs=0.548, $P<0.01$) across situations and significant differences were found in 'fearfulness' and 'affection-demanding' ($P<0.01$). There was little consistency between the caring staff's and experimenter's impression and the staff tended to evaluate the kittens as more active and affection-demanding but less aggressive than the experimenter. Differences were found in the degrees of 'fearfulness' and 'affection-demanding' across situations, suggesting a need for additional behavioural test in a novel situation to grasp the whole image of the kittens. The behavioral test used in this study would be helpful to provide more information.

The effect of visitors on behavior of goats at a petting zoo

Yoshitaka Deguchi[1], Mioto Yamada[1] and Takashi Iwase[2]
[1]*Iwate University, Faculty of Agriculture, 3-18-8 Ueda, Morioka, Iwate, 020-8550, Japan,* [2]*Morioka Zoological Park, 60-18 Shinjo-Shimoyagita, Morioka, Iwate, 020-0803, Japan;* *deguchi@iwate-u.ac.jp*

There is little research regarding the effect of visitors on the behavior of goats in a petting zoo. This study quantified visitor and goat behavior at a petting zoo in Morioka Zoological Park and examined a model for more favorable interactions. There were 20 female goats at the petting zoo and we observed 6 female goats that ranged from 1 to 9 years old. Observations occurred during two times of the year (Aug-Sep and Oct-Nov) and twice daily at 10:00-12:30 and 13:30-15:30. We recorded the behavior of the goats by continuous observation. When visitors interacted with the individual under observation, we recorded the behavior of the visitors, sex and approximate age. We did not communication to the visitors. The visitors allowed to feed of dead leaves to the goats in Oct-Nov. In addition, we recorded the number of people at the petting zoo every 15 minutes. The behavioral data was analyzed using one-way ANOVA with time period and the number of visitors as factors. A day was defined as a 'high visitor day' when the mean number of visitors was >5. A day was defined as a 'low visitor day' when the mean number of visitors was <5. The social investigative behavior of the goats to the visitors (watching, sniffing) significantly increased during high visitor days in Aug-Sep and Oct-Nov relative to low visitor days in Oct-Nov. The agonistic behavior of the goats to the visitors (head shaking, escape, avoid) significantly increased during high visitor days relative to low visitor days in Oct-Nov. The goats did not distinguish the visitors by age or sex. When the visitors touched the head and the back of the goats, 73% of the goats showed no reaction, 22% of them showed sniffing and 5% of them showed agonistic behavior. When the visitors demonstrated fast movements, such as chasing after the goats and trying to lift them, 80-100% of the goats responded with agonistic behavior. When the visitors offered dead leaves to the goats, 87% of the goats ate or smelled the leaves. Our study suggested that the behavior of the goats was influenced by the behavior of the visitors. When it was explained to the visitors that the goats disliked fast movements and being lifted, the visitors stopped doing these behaviors. Therefore, the goat welfare was improved and more positive visitor and goat interactions occurred.

Relationship between flight distance of mare and foal to human in Hokkaido native horse

Haruka Noda, Shingo Tada, Tomohiro Mitani, Koichiro Ueda and Seiji Kondo
Hokkaido University, Guraduate School of Agriculture, Kita 8, Nishi 5, Kita-ku, Sapporo Hokkaido, 060-0808, Japan; pandalucaaa@gmail.com

For aptitude of riding, characteristics of horses for easy acceptance against human are very important. Flight distance (FD) is known as index for acceptance. In a herd of mare-foal Hokkaido native horses, we hypothesized that mare with relatively low FD would have a foal with low FD. This study aimed to investigate acceptability against human of mare and foal. Experiment was conducted at Shizunai livestock farm of Hokkaido University from September to December 2014. Twenty-six pairs (mares and foals) of Hokkaido native horses kept outdoors all year around were used. FD of each mare and foal was measured by a laser distance meter. Distance between observer and horse was measured, then observer walked toward the horse and counted steps until the horse firstly started to escape. Measurements of FD were conducted consecutive 4 days in September and October, and consecutive 3 days for weaned foals in December. FD of mare and foal averaged with 2.8±2.2 m and 4.9±2.2 m. FD of mare and foal did not differ between September and October (mare; 2.8 vs 2.9 m, foal; 5.0 vs 4.7 m). There was significant positive correlation between FD of mare and foal (r=0.68, $P<0.01$). The correlation coefficient became lower when foals became elder (Sep; r=0.71, Oct; r=0.63). However, FD of weaned foals showed lower correlation coefficient in December (r=0.45, $P<0.05$). In conclusion, although the FD of foals correlated to FD of their mare, the relationship of FD of mare and foal weakened after weaning.

Effects of 1α-hydroxycholecalciferol and different phytase sources on broilers performance

Elmira Kabiri, Fatemeh Shirmohammad and Morteza Mehri
Islamic Azad University of Tehran, Animal Science, Shahid Kalhor Blvd, Shahre Qods, Fath highway, Tehran, 37541-374 Islamic Azad University of Shahre Qods, Iran; elmira.kabiri@yahoo.com

This experiment was conducted to investigate that the effect of two types phytase enzyme (microbial and fungal), 1α-OH-D_3 and different levels of calcium and phosphorus in diet on growth performance, carcass characteristics and bone traits of broiler chickens. 750 one day old broiler chicks (Ross 308) were distributed in a completely randomize design with six treatments and five replicates (25 birds in each replicate). Experimental treatments were included T_1: Control (without 1α-OH-D_3 and phytase), T_2: T_1+fungal phytase(50 g per ton), T_3: T_1+microbial phytase(50 g per ton), T_4: T_2+1α-OH-D_3(12 g per ton)+0.2% and 0.15% lower calcium and available phosphorus, T_5: T_3+1α-OH-D_3(12 g per ton) +0.2% and 0.15% lower calcium and available phosphorus, T_6: Control +1α-OH-D_3(12 g per ton), without D_3 and phytase. All experimental data was analyzed in accordance with the GLM Procedure. Differences among treatment means were determined using the Duncan's multiple range test. A probability level of $P<0.05$ was considered to be statistically significant. Results were shown experimental treatments had no significant effect on performance traits (Feeding behavior of broiler (feed intake, body weight gain and feed conversion ratio) was not affected by experimental treatmens). Also carcass traits and inner organ weight (except for liver) were not affected by experimental treatments. The relative weight of the Liver was significantly increased ($P<0.05$) in T3 treatment compared with T1 treatment (2.44 vs 2.08). Experimental treatments had no significant effect on bone traits and only Tibia Dyschondroplasia score in T_6 group was significantly lower (better) than other treatments ($P<0.05$). It seems decreased calcium and phosphorus level of diet did not have negative effect on performance and bone traits of broiler when the diet is supplemented with phytase and 1α-OH-D_3.

Effect of daytime and age on avoidance and approach behaviour of commercial danish broiler chicken

Franziska Hakansson and Helle H. Kristensen

University of Copenhagen, Department of Large Animal Science, Groennegaardsvej 8, 1870 Frederiksberg, Denmark; fh@sund.ku.dk

As activity levels of intensively managed broiler chicken vary over time, detailed knowledge on their influence is potentially useful to further adjust welfare assessment schemes. Therefore, this study investigated the influence of daytime and age on the performance of broiler chicken in applied fear tests. On-farm studies were carried out in eight flocks of intensively managed Ross-308 broiler chicken of three Danish producer. A forced and a voluntary approach test were conducted during morning and evening hours and at three different ages (1: 6-12 d; 2: 21-24 d; 3: prior to slaughter). At each observation, avoidance distances (AD) and the number of animals voluntarily approaching (VA) an observer were collected. Spearman correlation was used to investigate the effect of daytime on avoidance and approach, Mann-Whitney-test was used to look for differences in the distribution of the individual data sets. No significant correlation was found between daytime and AD or VA. When comparing data from morning and evening collected at different ages, significant difference could be found in the distribution of the VA at all ages (1: r=0.56*; 2: r=0.40*; 3: r=0.64*) but not for the AD. However, at three weeks the VA was found to be higher in the morning but at younger age and prior to slaughter, more birds approach an observer in the evening. Both, approach and avoidance peaked at three weeks with the latter decreasing to a minimum prior to slaughter. As activity and behaviour are known to vary over the day, detailed knowledge on their influence is potentially useful to further adjust welfare assessment schemes. From the results of this study I conclude, that the time of assessment has no effect on avoidance or approach behaviour of commercial broiler chicken. This makes the applied tests potentially applicable independent of time restrictions.

Research on animal-assisted intervention procedures for individuals with severe developmental disorders II

Toshihiro Kawazoe, Kazue Akabane, Takayuki Horii, Kanako Tomisawa and Itsuko Yamakawa
Yamazaki Gakuen University, Animal Nursing, 4-7-2 Minamiosawa Hachioji-shi, 192-0364
Tokyo, Japan; t-kawazoe@yamazaki.ac.jp

This research aimed to promote the spontaneous behavior of the individuals with severe developmental disorders by the interaction with dogs. They live in the care facilities, their actions need to be somewhat restricted. However repeated restrictions of their actions make them move only when they are told to, and they may restrict their movement by themselves. In those cases, since there is no spontaneous behavior, they might grow up without having the experience to learn how to give thought to others. Dogs have the possibilities to bring out the spontaneous behavior from the individuals with severe developmental disorders. However it is not necessary happened automatically by placing them with dogs. This research approaches to facilitate the spontaneous behavior of the individuals with severe developmental disorders with the effective positions and conditions of the dog. In this research we observed three case studies. The object individuals live in the support facilities for the disabled in the suburbs of Tokyo. The subject A has no physical disability, the subject B has attention deficit disorder, and the subject C is sometimes close to become unconscious by leaning to the left. Using behavior analysis, we researched the relations between 'the position of a dog' and 'the action of a subject'. We found that the effective position of the dog to facilitate the spontaneous behavior is different by each subjects. For subject A, the behavior to touch the dog was increased when dog gave a back comparing with when dog faced to the subject A. Subject B kept touching dog longer as same as the shaking of body axis was moderated when dog gave a back. For subject C, the time and frequency of watching dog were increased when the dog was located on the right. The spontaneous behavior was increased when we changed the position of the dog based on the disorders of the subjects. It is considered the subjects are praised for those behaviors repeatedly (reinforcement), this kind of experiences will increase the appropriate spontaneous behavior in the way of everyday life (generalization). We have recognized dog assisted therapy was effective by our activities in the past, and this time we found that the position of the subject and the dog is important to increase the spontaneous behavior of the individuals with severe developmental disorders. It is needed to indicate the relationship of the disorder characteristics and the position of the dog systematically by increasing the number of case studies. This research was conducted with authorization from the Yamazaki Gakuen University Ethics Committee.

Comparisons of behaviors between dog- and cat-assisted interventions in a nursing home

Naoko Koda[1], Yoshio Miyaji[2], Yanxia Song[1] and Chiemi Miyaji[2]
[1]Tokyo University of Agriculture and Technology, School of Agriculture, 3-5-8 Saiwai-cho, Fuchu, Tokyo, 183-8509, Japan, [2]Japan Animal-Assisted Therapy Council, 441-10 Shimokusadani, Inami, Kako-gun, Hyogo, 675-1101, Japan; koda@cc.tuat.ac.jp

Many animal-assisted intervention programs use dogs, and the use of other small animals is limited. However, cats are also a popular pet. This study compared behaviors between dog- and cat-assisted interventions in a nursing home. A visiting program for 14 residents with dementia was conducted 60 sessions in total. The residents were divided into two groups. Both groups interacted with both dogs and cats in counterbalanced order in 15 sessions each. Ten-minute interactions in each session were video-recorded and analyzed by counting the number of time blocks with a 10-second interval that the behaviors occurred. T-tests were performed. Both the dogs and cats stimulated the residents' interests and interactions with the visiting members. The dogs were more active (time blocks; dog: mean $\pm$ SD = 18.98$\pm$8.50, cat: 3.71$\pm$2.54, $P<0.05$) and made contact with the residents more frequently (dog: 3.84$\pm$3.22, cat: 0.55$\pm$0.54, $P<0.01$) than the cats. The residents watched the dogs for a longer time (dog: 15.12$\pm$8.19, cat: 12.99$\pm$7.19, $P<0.05$), and the handlers talked to the residents more frequently (dog: 4.48$\pm$1.69, cat: 3.49$\pm$1.08, $P<0.05$). Conversely, the residents more frequently petted the cats (dog: 3.06$\pm$3.73, cat: 7.15$\pm$7.64, $P<0.01$) and interacted with their handlers (dog: 1.42$\pm$2.00, cat: 4.83$\pm$5.11, $P<0.01$). During the cat interventions, the handlers more frequently approached (dog: 0.93$\pm$0.21, cat: 1.50$\pm$0.33, $P<0.001$) and left (dog: 0.64$\pm$0.26, cat: 0.99$\pm$0.36, $P<0.05$) the residents. The dogs were active while the cats were calm. The features of interactions were different according to the species presented. Cats can be used in such kind of programs for seniors as well as dogs.

Cats beg for food from the human who looks at and calls to them: ability to understand humans' attentional states in cats

Yuki Ito[1], Arii Watanabe[1], Saho Takagi[2], Minori Arahori[2], Mana Tsuzuki[2], Ayami Hyuga[2], Kazuo Fujita[2] and Atsuko Saito[1]
[1]The University of Tokyo, 3-8-1, Komaba, Meguro-ku, Tokyo, 153-8902, Japan, [2]Kyoto University, Yoshida-honmachi, Sakyo, Kyoto, 606-8501, Japan; assitu@cat.email.ne.jp

Recognizing humans' attentional states is important for companion animals to communicate with and form good relationships with their most important partners. Several studies have suggested that dogs in fact change their behavior depending upon attentional states or attentional foci of humans. In this experiment, we tested whether cats (Felis catus) can do the same in begging food from humans showing different attentional states. Thirty-three cats were given the occasion to beg food from one of the two unfamiliar experimenters (actors) sitting 1.5 m ahead, 1.5 m separated from each other and orienting toward the cat. Each experimenter held a piece of cat food in her hand on the chest. Almost all cats were tested in the following three conditions contrasting different attentional states. Condition A: the actor looking at and calling to the cat vs looking at the cat silently; condition B: looking at the cat silently vs face turned aside silently; condition C: facing down and calling to the cat vs facing down silently. Each condition included 2 warm-up trials and 4 test trials. The number of cats that passed the warm-up trials was 21, 21, and 22, in condition A, B and C, respectively. In each trial, we scored 1 if the cats correctly chose the actor providing more attentional information, -1 if they incorrectly chose the actor showing less attention and 0 if they failed to make a choice. The sum of the scores of 4 test trials was used for the analysis. The total score in condition A was significantly higher than 0 (chance level) (Wilcoxon signed-rank test: V=147, P<0.001). It shows that cats can use the vocal cue of being called when they beg food from humans. The total scores in conditions B and C were not different from 0. Thus the cats showed no preference for attentional states acted in conditions B and C. The total score in condition A was marginally higher than that in condition C (Wilcoxon signed-rank test: V=69.5, P=0.0501). The result in condition C shows that the same vocal cue did not work without actors' visual attention to them. These results suggest that cats can recognize humans' attentional state, by combining vocal and visual cues in begging food from humans.

Management of bumblebees to deliver biocontrol agents in open field conditions

Marika Mänd, Reet Karise, Riin Muljar and Gerit Dreyersdorff
Estonian University of Life Sciences, Institute of Agricultural and Environmental Sciences, Kreutzwaldi 1, 51014 Tartu, Estonia; marika.mand@emu.ee

Strawberry Fragaria × ananassa is a fruit crop grown worldwide, but diseases such as a grey mould Botrytis cinerea frequently limit yield. The majority of the grey mould infection on fruits originates from flowering period. Use of foraging bees as disseminators of microbial control agents (MCAs) to flowers is known as entomovector technology. Multiple researchers have shown that bumblebees can efficiently vector MCAs; however most of studies have been conducted under greenhouse conditions. The aim of this study was to investigate in what extent bumblebee Bombus terrestris visit strawberry flowers and whether it can suppress Botrytis cinerea in open field conditions where many competing plant species are flowering simultaneously. Field experiments were conducted in two strawberry fields (a´=0.5 ha) in Southern- Estonia (Rõhu and Polli) in 2012-2014. Bumblebee hives (Biobest Belgium NV) were placed near the fields (12 hives per ha) when the strawberry started flowering. Each hive had a special dispenser attached containing the biofungicide Prestop Mix with the spores of Gliocladium catenulatum J1446. Bumblebees flying out from the hive carried the biocontrol agent onto the strawberry flowers. The pollen pellets of homing bumble bees (n=960) were gathered and identified. To assess the effect of the bumblebee delivered Prestop Mix on grey mould infection rate the healthy and Botrytis-infected berries were counted. The Kruskal-Wallis ANOVA was used to test the effect of the year, location and treatment on the rate of strawberry infection by grey mold. All results were consided statistically significant at $P<0.05$, and all analyses were performed with the statistical package Statistica 12 (StatSoft, Inc). The study showed that bumblebee gathered pollen contained on average 22.4±1.3% strawberry pollen grains and 1/3 of them visited mostly strawberry during one foraging trip. The grey mould infection rate depended on the year (H (2, n=160)=69.54 $P<0.001$) and field location (H (1, n=160)=16.79; $P<0.001$). The rate of strawberry infection by grey mould was lower on the plots that were visited by bumblebees compared to isolated control plots: in 2012 65.3% ($P=0.007$); in 2013 37.3% ($P=0.043$) and in 2014 4.7% ($P=0.72$). This study provides strong evidence that bumblebees can vector a MCA to significantly reduce B. cinerea incidence not only in greenhouse strawberries but also in open field conditions where the landscape is heterogeneous with many competing flowers.

Finding the maximum slope for wild boars to climb up

Yusuke Eguchi[1], Hiroyuki Takeuchi[2], Soichiro Doyama[1], Hironori Ueda[1], Kenta Sakakura[2], Katsuji Uetake[2] and Toshio Tanaka[2]
[1]NARO Western region Agricultural Research Center, Yoshinaga 60 Kawai Oda Shimane, 694-0013, Japan, [2]Azabu University, 1-17-71 Fuchinome Chuou-ku Sagamihara-shi Kanagawa, 252-5201, Japan; eguchiy@affrc.go.jp

The purpose of this study is to measure and find the maximum angle of inclination for wild boars to able to climb up, in order to gain the basic knowledge and to develop useful techniques and strategies for the damage control. This experiment was carried out with two adult Japanese wild boars (one male, one female, 2 years old). We built the sloping apparatus that we can easily adjust the slope gradient. The length of slope was 2.7 m long. Two types of slope faces were used, one was made of wood chipboard with toeholds. The toeholds were made of wire mesh (mesh size: 10 mm and gauge size: 5 mm) and the other was the same size of wood chipboard without toeholds. Before the test, we placed the food as a reward on the slope with toeholds (at an angle of 10°) and on the goal at the end of the slope in order to let them learn how to get their rewards. For the tests, the reward was only placed on the goal. The angle of inclination was added by 5° each, starting from 20°. The duration of each trial was for maximum 5 minutes right after the boar got out of the box at the starting line until it climbed up the slope and got the reward at the goal. If they could not gain their rewards within 5 minutes, the trial was viewed as a failure. For the slope with toeholds, both boars were able to climb up at an angle of 45°. As they sometimes slipped down at this gradient, the test was discontinued for the safety reason. For the slope without toeholds, two boars were successfully able to climb up the slopes at angles of 20 and 25°, but at an angle of 30° both of them were failed. The result will be informative to find feasible methods for building fences in damaged hilly areas.

Investigation of observational learning ability of jungle crows

Christine Kamiya, Tsutomu Takeda, Masato Aoyama and Shoei Sugita
Utsunomiya university, agriculture, Tochigi prefecture, 321-8505 Utsunomiya city, Japan;
xkuragex_chr@yahoo.co.jp

Observational learning is the way of learning that animals learn by observing behavior of another one. There was a report that jungle crow could learn by observing another crow. Thus, the purpose of this study was to investigate whether jungle crow could learn by observing behavior of another crow that had done discrimination learning by color. Ten jungle crows were used in this experiment. These crows were divided into five pairs (demonstrator vs observer). The cage(250×300 cm) was divided in two by green net. Each one pair of crows was put in the cage, but they were not allowed to stay together because of the green net. Although they could see each other's appearance. At the first experiment, author used red and black paper for folding to cover feed tray. Feed were put only in tray covered with red paper. These two colors feed trays were put in front of demonstrator crow. When crow pecked one of two trays, these trays were changed to new trays and it was repeated ten times in a day. When crow could peck the correct color (red) with 90%, it is considered as succeeding in learning. At first demonstrator crow was made to learn the correct color (red), and after demonstrator crow could peck the correct color three days, author checked if observer crow could lean the correct color from observing behavior of demonstrator crow. In the second experiment, blue paper (correct color) and yellow paper were used and this experiment was carried out in the same way with first experiment. Except for the one exception, observer crow could learn faster than demonstrator crow and the correct rate of first day was higher than demonstrator crow. Thus, author concludes that jungle crow can learn by observing. But we don't know that to what extent jungle crow can learn by observing. So, we want to investigate the extent of observational learning of jungle crow.

Behavioral laterality in an Asian elephant (*Elephas maximus*) at Fukuyama Municipal Zoo, Japan

Shintaro Hagiwara[1], Naruki Morimura[2], Masayuki Tanaka[3], Chinobu Okamoto[4] and Shuichi Ito[4]
[1]Fukuyama Municipal Zoo, 276-1, Fukuda, Ashida-cho, Fukumaya-shi, Hiroshima-ken, 720-1264, Japan, [2]Kyoto University, Wildlife Research Center, Kumamoto Sanctuary, 990, Otao, Misumi-cho, Uki-shi, Kumamoto-ken, 869-3201, Japan, [3]Kyoto City Zoo, Okazaki Koen, Okazaki Houshoji-cho, Sakyo-ku, Kyoto-shi, Kyoto-fu, 606-8333, Japan, [4]Tokai University, School of Agriculture, Kawayo, Minamiaso-mura, Aso-gun, Kumamoto-ken, 869-1404, Japan; hagiwara_s@animbehav-tokai.com

Training is still the main means of Quality of Life improvement for elephants. In order to optimize training for better care management, we need to know the elephant's characteristics. Behavioral laterality, such as the dominant hand of humans, has been confirmed in all the other classes of vertebrates, but the details are not clear yet. It is assumed that better care management can be performed by taking behavioral laterality into account in training. In this study, we have investigated the sand-bathing, circling, and movements of the hind limbs while lying down and standing up of an Asian elephant. Our subject was an approximately 16-year-old female Borneo elephant at Fukuyama Municipal Zoo. For sand-bathing, during a total of 11 episodes, the act of flipping sand over the body was observed 126 times in total; over the left side of the body 49 times, over the right side 46 times, over the back side 20 times, and over the ventral side 11 times. The frequency of throwing on the left and right side was balanced. Lying down and standing up by command of the caretaker was conducted in 31 trials. For lying down, the elephant first put the right knee on the ground in all trials (chi-square test: $P<0.0001$). For standing up, the elephant first put the left hind limb on the ground significantly more ($P<0.05$). The elephant haven't been trained to movements of the limbs while lie down and stand up. All of the circling was clockwise walking. The foot of the elephant displayed clear signs of laterality of movements of the limbs. The nails were worn asymmetrically on the left and right side. In conclusion, behavioral laterality was observed in the movements of the hind limbs when lying down and standing up and circling. Behavioral laterality and body asymmetry should be taken into account in future.

Reaction of sika deer (*Cervus nippon*) to capsaicin repellent

Yuichiroh Shiiba and Ken-ichi Takeda
Shinshu University, Graduate school of Agriculture, 8304 Minami-minowa Kamiina Nagano, 399-4598, Japan; wrx-runboard@outlook.jp

The population of sika deer (*Cervus nippon*) in Japan has increased dramatically. Wire fencing has been used to prevent damage, but it's esthetically unsuitable for parks. Therefore, fence-less prevention methods are required. We investigated the effects of capsaicin repellent on five farmed sika deer. Three treatments were tested on each deer: the control fed from a box containing a 500 g hay cube (HC); in the repellent-smell treatment, 250 g repellent was placed at the bottom of a box and covered with wire under a 500 g HC and in the repellent treatment, 250 g repellent was mixed with a 500 g HC in the feeding box. Each treatment was investigated for 30 min. The experiments were conducted in the following order: control, repellent-smell, and repellent treatment. For each treatment, the amount of HC eaten, the time spent eating, and number of investigative behaviors were recorded. The three treatments were compared using the Steel-Dwass test. The mean amount of HC eaten differed for each treatment: control, 271.8 ± 105.9 g; repellent-smell treatment, 312.6 ± 53.6 g; and repellent treatment, 185.6 ± 61.7 g; ($P=0.07$). The mean time spent feeding on the HC was longer with the repellent treatments (repellent-smell treatment, 1071.0 ± 436.4 s; repellent treatment, 1030.0 ± 369.6 s) than the control (795.6 ± 529.9 s; $P=0.06$). The mean amount of exploration time was longer with the repellent treatments (repellent-smell treatment, 2.8 ± 3.3 s; repellent treatment, 37.3 ± 36.1 s) than the control (0.4 ± 0.8 s; $P<0.05$). Our results indicate that deer weren't affected by the smell of capsaicin, but food consumption decreased if it touched the mouth and mucous membranes of the deer.

The use of deer decoys in a corral trap to attract sika deer

Akitsu Tozawa[1] and Ken-ichi Takeda[2]
[1]Shunshu University, Faculty of Agriculture, 8304 Minami-Minowa Nagano, 399-4598, Japan,
[2]Shinshu University, Institute of Agriculture, Accademic Assembly, 8304 Minami-Minowa Nagano, 399-4598, Japan; akitsu@shinshu-u.ac.jp

Owing to the extinction of predators and the declining numbers and aging of hunters, the wild sika deer (*Cervus nippon*) population has considerably increased in Japan. Increasing deer populations have led to the damage of crops and forests. To prevent the damage and conserve natural resources, the effective capture methods that require little efforts and few people are anticipated. A corral trap can capture more than one animal at a time. A previous study has shown that deer decoys are useful for enticing deer into specific areas. Therefore, we experimentally evaluated the use of deer decoys in a corral trap to attract deer. Five captive female sika deer were used in the study. Two imitation corral traps (4 m^2, covered with silver plastic sheets 3 m-high, with a 2 m wide entrance) were set up next to each other in the rearing facility (the right corral trap: trap R, the left corral trap: trap L). The trap was divided diagonally into two parts: near the entrance and far from the entrance. Two decoys (one in a lying posture and the other in a grazing posture) were placed on one side of the trap (Treatment D), with no decoy on the other side (Treatment N). These decoys were placed on the trap R for 5 days, removed them from both traps for 3 days, and placed on the trap L for another 5 days. Video cameras were set up to record the number of deer entering the corral trap and the time they spent inside the trap. Data from both traps for the first three days were used for the analysis. Since no significant difference was observed between the data recorded for the trap L and R, both the datasets were summed up. The number of deer entering the trap (D: 4.3±2.2 times/head/day, N: 4.5±2.4 times/head/day, Binomial test *P*>0.05) and the time they spent inside the trap (D: 204.9±75.0 s, N: 168.9±42.3 s, t-test *P*>0.05) did not differ significantly between treatments. However, the correlation coefficients (r) of the time the deer spent inside the trap and the duration for which the deer had remained on the side far from the entrance were 0.75 and 0.48 in treatments D and N, respectively (*P*<0.05). These results suggest that decoys could be used as attracting materials for deer to remain inside a corral trap on the far side from the entrance.

Relationships among temperament, adrenal gland measures and productive performance in confined beef cattle

Aline C Sant'anna, Fernanda Macitelli Benez, Janaina S Braga and Mateus J. R. Paranhos Da Costa
Faculty of Agricultural and Veterinarian Sciences (FCAV), Animal Science, Unesp Campus, Jaboticabal-SP, 14883-298 Jaboticabal, Brazil; ac_santanna@yahoo.com.br

This study was an attempt to integrate cattle temperament with adrenal gland measures (as indicators of stress) and performance of feedlot beef cattle. The study was conducted using data from 97 Nellore and crossbred bulls (Angus × Nellore and Caracu × Nellore F1) raised on pasture and finished in a commercial feedlot, aged 30±6 months. During the study period (87 days) the animals were kept in outdoor feedlot pens with 6, 12 and 24 m^2 per bull. The temperament was assessed using two indicators: (1) the crush test recording the levels of cattle movements and tension inside the cattle crush, added in single reactivity scale (REA), ranging from 1 to 7; and (2) the flight speed test (FS, m/s) measuring the speed at which the animals exited the crush, faster animals were considered to have more excitable temperament. The finishing performance and carcass traits used were: final body weight (BW, kg), average daily gain (ADG, kg/day), hot carcass weight (HCW, kg) and ultimate meat pH (pH$_{24h}$). The right and left adrenal glands were collected during offal/viscera removal operation and, after dissection. They were taken to the laboratory, to obtain the following measures: ADCOR (cortical areas corrected by carcass weight), ADMED (medullar areas corrected by carcass weight) and ADWEIGT (total adrenal weight corrected by carcass weight). A principal component analysis was applied to the following variables: FS, REA, ADCOR, ADMED, ADWEIGT, BW, ADG, HCW and pH$_{24h}$. Four principal components (PC) had eigenvalues greater than one and, together, explained 70.3% of the variation in the data set. The PC1 had the highest positive loadings for BW (0.52), HCW (0.51) and ADG (0.43) and highest negative loadings for ADWEIGT (-0.31) and FS (-0.30), reflexing the positive association of FS with adrenal weight and negative with finishing performance. The PC2 and PC3 were not related to any performance trait and appeared to reflex the relationship between temperament traits and adrenal measures, since PC2 had higher positive loadings for ADMED (0.58) and ADWEIGT (0.53) and highest negative loading for FS (-0.28), while PC3 had highest positive loadings for ADCOR (0.51) and FS (0.38) and negative loadings for REA (0.64) and ADMED (0.31). The PC4 was the first component with high loading for meat pH$_{24h}$, with higher positive loadings for pH$_{24h}$ (0.71), REA (0.38), FS (0.35) and ADCOR (0.31). Our results indicate that cattle temperament is related to adrenal gland measures, with a trend for more excitable and flighty individuals (higher FS) to have adrenal glands with greater weight and cortical area, leading to worse finishing performance on feedlot and more elevated meat pH. Financial support: FAPESP (2013/20036-0).

Indicators of individual playfulness in rats

Jessica Lampe[1], Oliver Burman[2], Hanno Wuerbel[1] and Luca Melotti[1]
[1]University of Bern, Länggassstr. 120, 3012 Bern, Switzerland, [2]University of Lincoln, Riseholme Hall, Lincoln, LN2 2LG, United Kingdom; jessica.lampe@vetsuisse.unibe.ch

Play has been proposed as an indicator of positive emotions and welfare. The study investigated indicators of individual playfulness and its consistency across motivational contexts (with/without prior isolation) and time (2 time points). 24 adolescent male Lister Hooded rats housed in cages of 4 underwent 2 play tests: conspecific play in pairs (PIP) and heterospecific tickling (HT), which were compared to home cage play (HCP; first 2 h of dark-phase) and a novelty-induced suppression of feeding test assessing dependency of play and anxiety. For the play in pairs test, each rat was paired with each cage mate for 15 min, with and without prior 3.5 h isolation. The recorded play behaviour included solitary scampering (SC) and attacks to nape resulting in pinning (PN), partial rotation, no defense or evasion. For home cage play, PN was scored. In the 2-min tickling test, bouts of rapid finger movements on the rat's underside after 3.5 h isolation were performed. Positive ultrasonic vocalizations (USVs) and durations of hand-following behaviour (HF) were scored. The anxiety test measured latency to retrieve a food reward in the centre of an open arena. In play in pairs, frequencies of behaviours differed overall (all social play [SP] $F_{4,92}=76.0$, $P<0.01$), PN and SC being most frequent. Social play levels correlated across time within each motivational context (with isolation: $r=0.50$, $P=0.02$; without: $r=0.45$, $P=0.04$); PN and SC were consistent only with isolation ($r=0.50$, $P=0.02$; $r=0.59$, $P=0.01$). Play levels were related across motivational contexts at time 1 (T1; SP $r=0.46$, $P=0.02$; PN $r=0.54$, $P=0.01$; SC $r=0.68$, $P<0.001$), yet at time 2 (T2) only for SC ($r=0.62$, $P<0.001$). Consistency across time was found for USVs ($r=0.56$, $P=0.01$) but not hand following ($r=0.30$, $P=0.18$) and home cage play ($r=0.02$, $P=0.90$). Based on these results, play in pairs at T1 with isolation was compared with tickling test at T1 and home cage play, resulting in no correlations. The unrelatedness of play in pairs and home cage play was paralleled by differences in variances (PIP: 32.65, HCP: 118.95; $t(22)=3.28$, $P<0.01$), potentially making home cage play a less reliable measure. USVs and home cage play were unrelated ($r=-0.26$, $P=0.22$) as were USVs and HF ($r=0.14$, $P=0.52$). Anxiety and play were unrelated except for a tendency in USVs ($r=-0.40$, $P=0.06$) with least anxious rats vocalizing most. As play in pairs at T1 with isolation showed robust correlations across time and contexts, and USVs across time, they seemed to measure personality consistencies. However, as USVs had some relation to anxiety, the play in pairs test is proposed as the most valid indicator of individual playfulness in rats.

The significance of learning in homing of honey bee males

Shinya Hayashi, Mamoru Takata, Satoshi Koyama and Toshiyuki Sato
Tokyo University of Agriculture and Technology, Bioregulation and Biointeraction,
dongri22@gmail.com, 1900022, Japan; dongri22@gmail.com

Recently, learning impairment in honey bees is recognized as a serious problem for honey bee keeping. Honey bees are one of the major pollinators. Conservation of honey bee population is required to obtain sustainable agricultural production, including not only honey but also fruits. Understanding of the reproductive caste allows us to better conserve the population, although little is known about influence of learning on the performance of the reproductive caste, especially male honey bees. Honey bee males are known to take flight to mate, and return to their hive if they fail to mate with a queen. Males that failed to mate have to return to their hive to acquire subsequent reproductive opportunities because males are impossible to independently survive in the field. Therefore, homing ability plays an important role to increase reproductive success of males. Learning may be related to homing ability of males, because males need to memorize the position of their hive. Here, we investigated whether learning by males contributes to increase their homing performance. We compared the homing performance between males that could freely fly out and had opportunity to learn the position of their hive before experiment (learned group) and males that could not fly out and had no opportunity to learn (naive group). Both groups of males were released from a releasing point which is located at 200 m away from their hive. Furthermore, we investigated the effects of aging on homing performance of males by releasing 9, 12, and 15 days of age. To investigate the difference in homing rate between experimental groups, we used a generalized linear mixed effect model (GLMM) using R ver2.14.0. The homing rate was treated as a response variable assuming a binomial distribution, the experimental groups and age was treated as an explanatory variable, and origin of the male and date of experiment were treated as random factors. P-values were calculated using the likelihood ratio test. Males in the naive group did not return to their hive. 9 day old males in the learned group showed lower homing rate than 12 or 15 day old (9 vs 12 day: df=73, χ^2=6.8319, P=0.027; 9 vs 15 day: df=62, χ^2=7.0523, P=0.002; 12 vs 15 day: df=59, χ^2=0.2703, P=1.809). These results suggest that learning ability of males is essential to return their hive and would increase their reproductive success.

Effects of pasture experience on grazing behavior of sheep in a less-favored area

Yusuke Watanabe, Katsuji Uetake and Toshio Tanaka
Azabu University, Animal Science and Biothechnology, 1-17-71 Fuchinobe, Chuo-ku, Sagamihara-shi, 252-5201, Japan; ma1424@azabu-u.ac.jp

This study aimed to clarify the effects of pasture experience on the grazing behavior and the process of acclimatization of sheep in a less-favored area in Nagano Prefecture, Japan. In observation 1, six castrated Suffolk lambs without pasture experience were introduced to a formerly cultivated field (0.2 ha), and changes in grazing behavior and place use were observed for two months (August 4 – September 29, 2010). In observation 2, eight Suffolk ewes with or without pasture experience (four ewes each) were introduced to the same area, and changes in grazing behavior, place use, and distance between individuals were observed for two months (June 22 – August 21, 2013). In observation 1, all lambs stayed in the area near the entrance for most of the day following introduction. In contrast, all ewes used a wide area immediately after introduction in observation 2. It might be that ewes without pasture experience followed the ewes with pasture experience. In observation 2, differences in place use and individual distance by pasture experience were found immediately after introduction. The ewes with pasture experience used many areas significantly, and moved frequently and relatively long individual distances ($P<0.01$). At 19 days after introduction, although the ewes with pasture experience used a wider area than the ewes without pasture experience, there was no difference in the other two measurements, and a difference in all three measurements was no longer observed at 57 days after introduction. In conclusion, the sheep without pasture experience were judged to have acclimatized to the field about two months after introduction, and it was suggested that grazing a sheep herd that includes individuals with pasture experience in a formerly cultivated field is desirable. Both observations 1 and 2 were conducted during the summer season, but observation 1 was more than one month later. In addition, the sexes of the sheep were different. Thus, further studies on the effects of pasture experience on behavior are needed.

Questioning operant conditioning, from an ethical perspective

Francesco De Giorgio and José De Giorgio-Schoorl
Learning Animals | Institute for Zooanthropology, Achterstraat 64, 5833 TP Nistelrode, the
Netherlands; info@learning-animals.org

It has been many years now, that the behaviorist paradigm, both in its interpretative models, as in its various forms of application as in the case of operant conditioning, is a discussed Cartesian-like paradigm from a bio-naturalistic and cognitive point of view. Still, with the ongoing study of the human-animal relationship as exponentially growing field of attention, and with more scientific insight and data about the emotional-cognitive side in animals, as well as the understanding of the side effects of overtraining when studying behaviour in a scientific context, the trend of applying operant conditioning in day-to-day practice, focussing on stimulus-response protocols seems to become the predominant language and approach when interacting with non-human and human animals. This tendency is expanded towards human animals as well in order to work towards a desired behavioural expression as result. But is it really about the result? Should it be about a behavioral expression? And what if these results lead to side effects that have an impact on animal minds? And even if there are results, with no apparent side effects, should we still consider it an acceptable paradigm, by a new ethical perspective? Interfering with their possibility to create their own dialogue with their surroundings, and decide what information to find interesting? Where does it leave the animal as sentient being? This brings us to the other side of this case; are there alternatives to the use of operant conditioning? These questions are intended to moving on to a new ethical paradigm and practical application, following more real and innovative cognitive models of interaction, to give value to individuality, socio-cognitive abilities, subjectivity and intrinsec value of non-human animals.

Effect of illuminance on hens' colour discrimination abilities

Yuko Suenaga[1], Yumi Nozaki[1], Shuho Hori[2], Chinobu Okamoto[1], Ken-ichi Yayou[3] and Shuichi Ito[1]

[1]*Tokai University, School of Agriculture, Kawayo, Minamiaso-mura, Aso-gun, 869-1404,Kumamoto-ken, Japan,* [2]*Tokai University, Graduate School of Agriculture, Kawayo, Minamiaso-mura, Aso-gun, 869-1404,Kumamoto-ken, Japan,* [3]*National Institute of Agrobiological Sciences, Tsukuba-shi, 305-8602,Ibaraki-ken, Japan; yscandy1205118@yahoo.co.jp*

It has been believed that poultry can see colours; however, this capability is thought to be affected by the illuminance level, such that hens are less capable of discriminating colours in dark environments than are other birds or mammals. In this study, we clarified the colour discrimination ability of hens under low light conditions. We raised four birds each of the Julia (a~d) and Boris Brown (A~D) varieties. We trained the hens to learn the relationship between a positive target (a red coloured card) and a reward (feed) and to discriminate between red and grey cards. We used 10 red cards as a positive stimulus and 10 grey cards as a negative stimulus, which were random brightness. Therefore, the subjects could not choice a positive stimulus to clue the brightness. The left and right positions of the two cards were shifted at random. Each session consisted of 20 trials. The criterion of successful discrimination was two consecutive sessions with more than 16 correct choices ($P<0.01$, Chi-square test). After the hens were fully trained, we tested their colour perception abilities in low light intensities. In the training session, only two birds learned the relationship between positive stimuli and reward. Two Boris Brown birds (Birds [C] and [D]) learned to discriminate the coloured cards under normal light intensities. Bird [C] required 4 sessions and Bird [D] required 11 sessions to reach the criterion. Once the criterion was reached, the illuminance of light was gradually reduced. Both individuals were able to distinguish at 1 to 66 lx. In addition, Bird [C] was able to distinguish at down to 0.1 lx. The results of this study suggest that the colour vision of hens is well developed and that at least some hens are able to discriminate coloured objects under low light conditions.

Training of tool-use behavior in rats

Akane Nagano[1] and Kenjiro Aoyama[2]
[1]Doshisha University, Graduate School of Psychology, 1-3 Tatara Miyakodani, Kyotanabe-shi, Kyoto, 610-0394, Japan, [2]Doshisha University, Faculty of Psychology, 1-3 Tatara Miyakodani, Kyotanabe-shi, Kyoto, 610-0394, Japan; ekp1003@mail2.doshisha.ac.jp

Tool-use behavior has been observed in non-human animal species such as keas and common marmosets. Tool use has been observed in the wild and in experimental settings. A species of rodent (degu) has been previously used as animal models to investigate the tool-use behavior. In the present study, we showed that rats can be trained to use tools to obtain food in an experimental setting. We used four experimentally naïve Brown-Norway rats (animal number: BN1-BN4), approximately 2 months old at the start of the experiment. During the training and testing phase, all rats were maintained at approximately 90% of their initial free feeding weight. The rats were trained to use the rake-shaped tool to retrieve a food (one eighth of chocolate flavored loop cereal) beyond the rats' reach in an experimental chamber. Each daily experimental session consisted of 40 trials. Training in the each phase continued until the rats reached a criterion of 30 or more success trials in three consecutive sessions. The success trials were recorded if the rat could obtain the food within one minute. In the first phase, the food was placed between the blade of the tool and the rat. In this phase, the rats learned to retrieve the food simply by pulling the tool directly towards them. In the second phase, we placed the food at the side of the tool so that, to retrieve the food, the rat had to move the tool laterally before pulling it. The tool and the food were placed on the left side or right side of the experimental chamber. When the tool was placed on the left side of the chamber, the food was placed on the tool's right. When the tool was placed on the right side of the chamber, the food was placed on the tool's left. In the final phase, the position of the tool and the food was changed to test whether rats could move the tool according to the position of the food. The tool was placed in the center of the experimental chamber, and the food was placed on the tool's left or right. The position of the food was pseudorandomized. We analyzed the direction of the tool manipulation on the first day of the final phase. A chi-square test revealed that three of four rats moved the tool according to the position of the food on the first day of this phase (BN1: χ^2 (1)=14.30, $P<0.001$; BN2: χ^2 (1)=10.53, $P<0.001$; BN3: χ^2 (1)=0.95, n. s.; BN4: χ^2 (1)=19.60, $P<0.001$). The results indicate that rats moved the tool according to the position of the food.

Aboveground behavior of plateau pika (*Ochotona cruzoniae*) in the grazing alpine rangeland in the Tibetan Plateau

Nobumi Hasegawa[1], Jiahua Yang[2], Masahiro Tasumi[1], Akira Fukuda[3], Sachiko Idota[1], Manabu Tobisa[1], Guomei Li[4] and Rende Song[5]
[1]University of Miyazaki, Miyazaki, 889-2192, Japan, [2]Qinghai University, Xining, Qinghai, 810016, China, P.R., [3]Shizuoka University, Hamamatsu, 432-8561, Japan, [4]Yushu Prairie Center, Yushu, Qinghai, 815000, China, P.R., [5]Yushu Animal Husbandry and Veterinary Center, Yushu, Qinghai, 815000, China, P.R.; nhasegawa1411@gmail.com

Plateau pika (*Ochotona cruzoniae*) is a small mammal which inhabits the alpine rangeland in the Tibetan Plateau. It burrows underground and digs up a large amount of soil which covers the vegetation and produces bare patches there. It also forages plants and stores them in the burrows for its winter-season diet. It has been poisoned as a pest animal which detrimentally affect the vegetation of the rangeland. However, it is considered that it plays an important role as a keystone species for the biodiversity and the circulation of energy and materials in the ecosystem. In order to obtain the scientific and fundamental knowledge for the proper management of the rangeland and the preservation of the ecosystem, aboveground behavior of a plateau pika was observed using a scouting camera for nine consecutive days in August of 2012 in the rangeland in Yushu, Qinghai, China. Data was statistically analyzed by ANOVA and Tukey-Kramer HSD test. Aboveground active bout of plateau pika was from 6:00 to 20:59 and was not before 6:00 and after 21:00. It was divided into five 3-hour Periods (I: 6:00:00-8:59:59, II: 9:00:00-11:59:59, III: 12:00:00-14:59:59, IV: 15:00:00-17:59:59 and V: 18:00:00-20:59:59). It was significantly different among the Periods ($P<0.001$), and was significantly higher in III ($34.4\pm15.9\%$) and lower in I ($3.7\pm5.0\%$) than in the others ($P<0.05$). When the behavior was classified into four categories, the rate was significantly greater in keeping still and/or watching ($54.8\pm14.6\%$) than in the others (foraging, $20.7\pm8.1\%$; locomotion, $10.4\pm3.6\%$; and other behavior, $14.1\pm5.2\%$) ($P<0.05$). Raptores (*Buteo hemilasius* and *Athene noctua*) and carnivores (*Vulpes ferrilata* and *Mustela altaica*) were observed around the burrows of plateau pikas. It is considered that the reason why the plateau pika behaved at high rate of keeping still and/or watching was to be cautious of predators. This work was supported by JSPS KAKENHI Grant Number 23255015.

Structure of subgroup in mares and foals in a herd of reproductive horse and formation change of subgroup in weaned foals

Fumie Sato[1], Singo Tada[2], Tomohiro Mitani[1], Koichiro Ueda[1] and Seiji Kondo[1]
[1]Hokkaido University, Graduate School of Agriculture, Kita-ku, Kita 9, Nishi 9, Sapporo, 060-8589. Hokkaido, Japan, [2]NARO Hokkaido Agricultural Research Center (NARO/HARC), Division of Animal Science, 1, Hitsujigaoka, Toyohira-ku, Sapporo, 062-8555, Japan; tdshng@affrc.go.jp

As a follower-type offspring-care animals, a mare and her foal would be usually positioning together spatially. If there were sub-groups (SG) of mares in a herd, SG of foals would exist with SG in mares. In this study, we hypothesize: (1) there are SG of mares and SG of their foals would be positioning together, (2) members in SG of foals would change after weaning. In a herd of Hokkaido native ponies kept outdoors all year around, 28 pairs of mare-foal were observed in 7 days at a 1-hour intervals through daytime in October (pre-weaning), then the same observation was done on herd of 28 foals after weaning in December. In each observation, animals existing within 3-body length from every animal were identified and recorded. According to proximal frequencies of appearance for each animal, social network were made by the software of Pajek64 (ver.3.11) and SG for mares and foals were respectively determined by the method of Louvain. Four SG of mares and 4 SG of suckling foals were detected in the networks and the members of each SG were almost the mares and her foals. In the group of foals after weaning, 5 SG were detected. In these SG of weaned foals, 4 SG were formed partially from the members of the SG in pre-weaning. It was suggested that mare-foal relationship in such a maternal herd of reproductive horses would be continued through generations.

Improved tunnel ventilation of dairy cows in a tie-stall and face-out type barn

Mizuna Ogino[1], Koshin Haga[1], Kazuna Nishikawa[1], Risako Mizukami[1] and Kazuhiko Higashi[2]
[1]Ishikawa Prefectural University, Bioproduction Science, Nonoichi, 921-8836, Ishikawa, Japan,
[2]Ishikawa Prefecture, Kahoku, 929-0325, Ishikawa, Japan; ogino@ishikawa-pu.ac.jp

Livestock with high levels of metabolizable energy are significantly affected by heat stress during summer, and in cows, this can lead to a decrease of milk yield and productivity. Recently, tunnel ventilation has been introduced to improve heat stress and is characterized by air inlets at one end and exhaust fans at the other. These features generate a stronger flow of air that flows in a straight line through the barn, and it is possible that tunnel ventilation is more effective than other ventilation systems. In this study, we improved tunnel ventilation in a tie-stall and face-out type barn, which is the modification that separated the air into left- and right-sided flows, and investigated the effect on dairy cows by measuring behavior, heart rate variability, and blood stress indicators. Dairy cows were maintained in a tie-stall and face-out type barn in Kahokugata, Ishikawa prefecture, Japan. The improvement which involved setting the partition in the inlet of the barn, was enforced in all experiments except the first. The behavior of all cows (n=60) was observed at 10-min intervals, and environmental measurements (temperature, humidity, and wind velocity) were taken at 30-min intervals at nine points in the barn. Behaviors were recorded with postures. Blood was collected and the heart rate variability, body surface temperature, and rectal temperature of lactating cows at six points (three per side) in the barn were measured. The temperature-humidity index (THI) and mean body temperatures were calculated. Since mean body temperature increased with increasing THI (r=0.68, $P<0.001$), the thermal environment of the barn affected mean body temperature. The amount of rest spent in a standing posture increased with increasing THI (r=0.35, $P<0.05$) and that on a lying posture decreased with increasing THI (r=-0.43, $P<0.05$). Plasma cortisol concentration was the highest in July when THI was the highest ($P<0.05$). However, there was no relationship between wind velocity and other environmental or physiological variables measured. The normal mean body temperature in cows is 38.1-38.7 °C. Since the mean body temperature of all experimental cows was normal, this suggests that the heat stress did not strongly affect the cows and, in this study, the effect of improved tunnel ventilation was not clear.

Effect of outside temperature on utilization of shelter in a pasture for young Holstein heifers in early spring

Yuko Shingu, Yukiko Nishimiti and Ikuo Osaka
Konsen Agricultural Experiment Station, Nakashibetsu-cho Hokkaido, 086-1135, Japan;
singuu-yuuko@hro.or.jp

In early spring, setting a shelter to maintain warmth in a pasture may modify the effects of drastic temperature change. The purpose of this study was to determine the effects of outside temperature on the utilization time of shelter for first-grazing-season heifers. Three Holstein heifers (mean BW 193 kg) and four Holstein steers (mean BW 218 kg) were kept for 31 days in early spring on a pasture (0.45 ha) containing a plastic greenhouse (158 m^2). On days 1, 2, 8, 9, 15, 16, 22, 23, 28, and 29, the number of recorded positions on the pasture area or the plastic greenhouse area was determined by recording three heifers' positions at 15-sec intervals with a GPS receiver. Using the number of positions, the utilization time in each area at each outside temperature was calculated during behaviour except for grazing and drinking only on observation days in which over 70% of the positions could be recorded by the GPS receiver (n=18). Outside temperature was measured at 15-min intervals. Mean daily outside temperature and wind speed were 13.3 (SD±4.4)°C and 2.8 (SD±1.4) m/s. Mean utilization time of the plastic greenhouse on all observation days (78 min/day) was significantly shorter than that of the pasture (733 min/day) (GLM procedure, $P<0.05$). Daily mean utilization time in the plastic greenhouse at 0-5 °C (103 min, n=2) was significantly longer than those at 5-10 °C (14 min, n=9), 15-20 °C (11 min, n=13), 20-25 °C (4 min, n=9), and 25-30 °C (10 min, n=6) (Tukey-kramer HSD test, $P<0.05$). First-grazing-season young heifers tend to favour utilization of the plastic greenhouse, indicating that a shelter should be set in a pasture in early spring, particularly when the outside temperature falls to 0-5 °C.

Effects of water restriction following feeding during heat stress on behavior and physiological parameters of Corriedale ewes

Jalil Ghassemi Nejad, Byong-Wan Kim, Bae-Hun Lee, Do-Hyeon Ji, Jing-Lun Peng and Kyung-Il Sung
Kangwon National University, Animal Life System, Lab no 302. College of Animal Life Sciences, 200-701, Korea, South; jalilghaseminejad@yahoo.com

This study was conducted to investigate the effect of water restriction following feeding on behavior and physiological parameters of heat stressed Corriedale ewes. Nine Corriedale ewes (average BW=43±6.5 kg) were individually fed diets based on maintenance requirements in metabolic crates. Sheep were allotted to three groups according to a 3×3 Latin square design for 3 periods with 21-d duration for each period (two months, 9 sheep per treatment). Sheep were acclimated to the environment and experimental house conditions for 10 d prior to the experiment. Average temperature-humidity index (THI) in the housing unit (>27) defined heat stress condition. Statistical analysis was carried out using the GLM procedure of SAS. Treatments were free access to water (FAW), 2 h water restriction (2hWR), and 3 h water restriction (3hWR) following feeding. No differences were found in fecal excretion frequency (3.7, 4.0, and 4.2 no./day for FAW, 2hWR, and 3hWR group, respectively, SEM=0.3), standing times frequency (19.1, 20.8, and 22.3 no./day for FAW, 2hWR, and 3hWR group, respectively, SEM=1.4) and sitting times frequency (18.2, 19.8, and 20.8 no./day for FAW, 2hWR, and 3hWR group, respectively, SEM=2.1) among treatment groups ($P<0.05$). Measures of standing duration (317.9, 275.2, and 273.4 min/d for FAW, 2hWR, and 3hWR group, respectively, SEM=17.4) and urine excretion frequency (10.4, 6.9, and 6.9 no./day for FAW, 2hWR, and 3hWR group, respectively, SEM=1.2) showed significant decrease whereas sitting duration (405.1, 446.0, and 449.4 min/d for FAW, 2hWR, and 3hWR group, respectively, SEM=17.7) showed significant increase in 2hWR and 3hWR groups compared with FAW group ($P<0.05$). Fecal score (1.3, 1.3, and 1.2 for FAW, 2hWR, and 3hWR group, respectively, SEM=0.1) and heart rate (66.0, 66.0, and 67.5 no./min for FAW, 2hWR, and 3hWR group, respectively, SEM=1.2) were not different among treatment groups ($P>0.05$). However, respiratory rate (88.8, 107.2, and 106.1 no./min for FAW, 2hWR, and 3hWR group, respectively, SEM=2.9) and panting score (1.5, 1.9, and 1.9 for FAW, 2hWR, and 3hWR group, respectively, SEM=0.1) were significantly higher in 2hWR and 3hWR groups than FAW group ($P<0.05$). It is concluded that water restriction following feeding could intensify physiological heat stress related indicators and change behavioral parameters in heat stressed ewes. Sheep showed daily adaptation effectively from morning to evening to hot and humid environment followed by water restriction.

The effects of parity on sexual behaviours in group weaned sows

Rebecca Woodhouse[1], Rebecca Morrison[2], Paul Hemsworth[1], Jean-Loup Rault[1], Christian Hansen[3] and Lisbeth Hansen[4]

[1]*University of Melbourne, Animal Welfare Science Centre, Alice Hoy Building, Parkville VIC 3052, Australia,* [2]*Rivalea Australia, Corowa, NSW, 2646, Australia,* [3]*University of Copenhagen, Department of Large Animal Sciences, Groennegaardsvej 2, 1870 Frederiksberg C, 1017 Copenhagen K, Denmark,* [4] *Danish Agriculture and Food Council, Pig Research Centre, 1609 København, Copenhagen, Denmark; rebecca.woodhouse@unimelb.edu.au*

We investigated the effects of parity on sexual behaviour of sows housed in groups post-weaning. At weaning, 180 sows were housed in pens of 10 at 4.4 m^2 per sow, with 6 groups of parity 1 and 3 groups each of parities 3, 4, 5, and 6. Sexual behaviour was analysed from days 2 to 5 using one-zero sampling during 5 min intervals every 30 min from 07:30-17:30 h. Continuous data were analysed using a mixed model and categorical data using Chi-square or Fisher's test. Wean-to-mate interval did not differ between parities ($P=0.46$), but the overall frequency of sexual behaviour initiated between days 3 and 6 differed, with parity 1 initiating less sexual behaviour than parity 6 (14.9 ± 3.1 vs 31.7 ± 4.6 bouts, $P=0.03$). Parity 1 also received less sexual behaviour (12.0 ± 3.4 bouts) than parities 4 and 5 (28.7 ± 3.9 bouts, $P=0.02$; 36.7 ± 4.1 bouts, $P<0.0001$, respectively). Specific sexual behaviours differed according to parity, with more mounts for parity 6 than parity 3 (2.9 ± 0.6 vs 0.6 ± 0.6 bouts, $P=0.04$). Ano-genital sniffing also differed according to parity ($P<0.0001$), with more sniffing in parities 4 and 6 compared to parity 1 (19.0 ± 2.6 bouts and 24.8 ± 2.7 vs 8.2 ± 1.9 bouts, $P=0.008$ and $P<0.001$, respectively). Parity is not always considered when grouping sows in commercial settings. Lower parity sows may be at risk of receiving potentially injurious or stressful behaviours such as mounting from more sexually active higher parity sows. Lower parity sows are also not as likely to display sexual behaviour when in oestrus, suggesting that more accurate oestrous detection methods may be required for lower parity sows in mixed parity groups.

A gradual reduction in sow contact prepares piglets for weaning

Emily De Ruyter[1,2], David Lines[1,3], William Van Wettere[1,2] and Kate Plush[1,2,3,4]
[1]CRC for High Integrity Australian Pork, Davies Building, Roseworthy SA, 5371, Australia, [2]The University of Adelaide, School of Animal and Veterinary Science, Davies Building, Roseworthy SA, 5371, Australia, [3]Australian Pork Farms Group Pty Ltd, Stirling, 5152, Australia, [4]South Australian Research and Development Institute, Livestock Farming Systems, Davies Building, Roseworthy SA 5371, 5371, Australia; kate.plush@sa.gov.au

This study tested whether a gradual reduction in sow contact during lactation influenced piglet stress response to weaning. Sow contact was reduced by separating the sow from piglets (SP, n=30) for 5 h per day from days 10-15 of lactation, 7 h on days 15-20, and 9 h on days 20-weaning. Litters from 20 sows were followed as controls (CON), remaining in full contact with one another until weaning. Weaning occurred on day 28±1 of lactation. Piglet weight and injury scores were measured throughout lactation and after weaning. Continuous video footage was collected from 07:00 to 13:00 on the two days following weaning for behavioural analyses. Piglets were bled the day prior to and the day following weaning, and change in plasma cortisol concentration was calculated from these samples. All not normally distributed data were transformed prior to analysis, conducted in SPSS. Piglet weight, injury score and behaviour following weaning were analysed using a linear mixed model, and change in cortisol concentration using a general linear model. SP piglets were lighter than CON piglets at weaning (CON 7.6±0.2 kg, SP 6.8±0.2 kg) but this weight disparity had disappeared by day 7 post weaning (CON 8.6±0.2 kg, SP 8.4±0.1 kg, $P>0.05$). Belly nosing and aggressive events were longer in duration for CON piglets on the day following weaning (belly nosing: CON 6.3±2.0 sec and SP 2.4±1.3 sec; aggression: CON 6.5±1.1 sec; SP 4.2±0.8 sec; $P<0.05$). Injury scores were higher for CON piglets on almost all days examined ($P<0.05$). An increase in circulating plasma cortisol concentration in response to weaning was observed in CON piglets (18.7±13.3 nmol/l), but a negligible change was identified in the SP piglets (-12.3 ±14.1 nmol/l; $P<0.05$). These findings imply sow separation during lactation provides welfare benefits for piglets around the highly stressful weaning period.

Isolation affects locomotive behavior and food intake in Japanese Himedaka

Eri Iwata and Koutarou Sakamoto
Iwaki Meisei University, Department of Science and Engineering, 5-5-1, Chuoudai, Iino, Iwaki City, Fukushima Prefecture, 970-8551, Japan; asealion@iwakimu.ac.jp

It is well known that environmental conditions have a major effect on individual behavior in mammals. Recently, many studies have found that this phenomenon also exists in other vertebrates, such as reptiles, amphibians and teleosts. In teleosts, the environmental effects on hatchery fish behavior have been investigated because of the industrial requirement, but the effects on ornamental fish are not well documented. The Himedaka is one of the traditional Japanese ornamental fish bred from wild Japanese killifish Oryzias latipes, which live in schools, and also known as a model organism for many areas of biological research. Then we compared the behavior of isolated- and group-housed adult Himedaka to evaluate the environmental effects on fish behavior in small fish tanks, which were generally used for private aquariums or laboratory experiments. Adult fish which had been kept in a semi-natural pond were introduced into laboratory conditions and kept for 30 days for habituation. Then, the fish were divided into isolated- and group-housed experimental groups, and further kept for 30 days. On the 30th day, the fish were videotaped periodically during daytime hours, and the duration % of resting for the entire observation time was calculated. After the behavioral observation, the total body length and mass of the fish were measured. Irrespective of sex, the isolated fish exhibited the longer duration % of resting compared to the group-housed fish, while the group-housed fish were almost always in locomotion during the observation time ($41.61\pm9.78\%$ for isolated and $4.15\pm1.02\%$ for group-housed fish; Mann-Whitney U test, $P<0.05$). The isolated fish showed a reduction in body mass compared to the group-housed fish (0.75 ± 0.01 g for isolated and 0.56 ± 0.02 g for group-housed fish; Two-way ANOVA and Tukey-Kramer, $P<0.05$), though the total body length showed no significant difference. These results suggest that environmental conditions may have a great impact on fish behavior and health conditions as well as in mammals. Therefore, providing proper housing conditions are expected to contribute to fish welfare.

Effect of season and existence of female on crowing in red jungle fowl

Shuho Hori[1], Mari Iwahara[2], Makiko Hirose[2], Atsushi Matsumoto[3], Masayuki Tanaka[4], Tsuyoshi Shimmura[5] and Shuichi Ito[2]

[1]Tokai University, Graduate School of Agriculture, Kawayo, Minamiaso-mura, Aso-gun, Kumamoto-ken, 869-1404, Japan, [2]Tokai University, School of Agriculture, Tokai University, School of Agriculture, Kawayo, Minamiaso-mura, Aso-gun, Kumamoto-ken, 869-1404, Japan, [3]Kumamoto City Zoological and Botanical Gardens, 5-14-2, Kengun, Higashi-ku, Kumamoto-shi, Kumamoto-ken, 862-0911, Japan, [4]Kyoto City Zoo, Okazaki Koen, Okazaki Houshojicho, Sakyo-ku, Kyoto-shi, 606- 8333, Japan, [5]National Institute for Basic Biology, 38 Nishigonaka, Myodaiji, Okazaki-shi, Aichi- ken, 444-8585, Japan; tetra62@gmail.com

Crowing, the characteristic call of roosters, is controlled by a circadian clock. However, most of its behavioural functions remain unproven. Here, we report two observations on the seasonal change of crowing (Exp. 1) and the effect of presence of female on crowing (Exp. 2) in red jungle fowls. In Exp. 1, we observed the seasonal change of crowing in a group (four males and three females) under semi-wild conditions in a zoo. The Crowing of males was the continuously recorded 24 h for 2 days in spring (breeding season) and in autumn (non-breeding season) by IC recorder. The individual crowing of the red jungle fowls could easily be distinguished by the sound waveforms using iMovie. In Exp. 2, we observed four males and one female introduced into individual cages in 25 January 2015 under 12-h light: 12-h dim light cycles for 14 days. Males were kept in the same room with the female for the first 7 days, and then the female was removed from the room for the next 7 days. In Exp. 1, all males crowed at twilight before sunrise. While the first crowing was observed approximately 137 (±2) min before sunrise in spring, the crowing began approximately 75 (±23) min before sunrise in autumn. In addition, crowing was observed more frequently in spring than in autumn (number of crows 352±107 vs 87±63, $P<0.05$, paired t-test). Individual differences were also observed in spring, whereas differences were small in autumn. In Exp. 2, we focused on two males demonstrating crowing. The number of crows of these two males increased after removal of the female from the room (282±35 vs 398±18 and 229±22 vs 274±36 crows, respectively). These results suggest that the crowing of red jungle fowls may be affected by season and presence of female.

Low space at mixing negatively affects low ranked sows, according to a novel method for analysis of sow social structure

Emma C Greenwood[1], Kate J Plush[2], William HEJ Van Wettere[1] and Paul E Hughes[2]
[1]University of Adelaide, School of Animal and Veterinary Science, Roseworthy campus, Roseworthy, SA, 5371, Australia, [2]South Australian Research and Development Institute, Roseworthy, SA, 5371, Australia; emma.greenwood@adelaide.edu.au

To date, calculations for social rank have involved fights won or lost on the day of mixing, or successful displacements over a trial period, but not both. The aim of this investigation was to determine if a relationship between space at mixing and social rank existed using a new method of dominance classification. A total of 132 sows were mixed into groups of six after insemination, with a space allowance of 2, 4 or 6 m^2/sow. Behaviour over six hours was recorded at mixing (day 0) and days 1, 3 and 4. Salivary cortisol and injuries were also recorded. 'Hierarchy' was calculated using the number of fights won and lost on the day of mixing in conjunction with the number of successful displacements over the entire experimental period. Data was analysed using a linear mixed model in SPSS (transformed data is presented with untransformed data in brackets). 1D-no displacements 2D-displaced more often than displaced others 3D-displaced others more often than displaced 1F-no fights 2F-lost more fights than won 3F-won more fights than lost Although 1D1F sows were involved in no fights on day 0 they received more aggression (bites received 1D1F=0.7±0.1 (5.8), average for remaining classes=0.4±0.1 (1.9), $P=0.041$) and had more injuries (1D1F=6.36±0.43 (55.16), average for remaining classes=4.69±0.32 (27.2), $P<0.005$) than all other groups. 1D1F sows had more injuries when kept in 2 m^2/sow allowance compared with other space allowances (2 m^2/sow=8.5±1.1 (71.6), 4 m^2/sow=5.6±0.9 (31.9), 6 m^2/sow 5.0±0.6 (24.9), $P=0.008$). Analysis using both fights and displacements revealed previously unseen differences between subgroups of sows. For example, 1D1F sows faced more aggression than others, suggesting they were targets for aggression but did not retaliate, unlike 1D2F sows. It is evident that the provision of higher space allowances at the point of mixing provides welfare benefits for low ranking sows. This project was funded by the Pork CRC and the University of Adelaide.

Reducing sow confinement during parturition and early lactation has no effect on sow maternal behaviour in later lactation

Patricia C Condous[1], Kate J Plush[2] and William HEJ Van Wettere[1]
[1]University of Adelaide, School of Animal and Veterinary Science, Roseworthy, 5371 SA, Australia,
[2]South Australian Research and Development Institute, Roseworthy, 5371 SA, Australia;
patricia.condous@adelaide.edu.au

Housing systems which confine sows only during parturition and early lactation represent an alternative to farrowing crates. However, there is little knowledge of how partial confinement affects maternal behaviour. This study aimed to determine the effect of sow confinement and non-confinement during parturition and early lactation on sow behaviour in later lactation. Twenty-six sows and litters were housed in a conventional farrowing crate (1.7×2.4 m) or a swing-sided pen (SS pen) (2.8×2.2 m). Four treatments were created within the SS pen: (1) open during parturition and re-opened day 3 lactation; (2) open during parturition and re-opened day 7 lactation; (3) closed during parturition and re-opened day 3 lactation; (4) closed during parturition and re-opened day 7 lactation. Sows were recorded for 5 h on day 8 of lactation and the day before weaning (day 26 ± 2 of lactation). Time taken to lie down, sow-piglet interactions and nursing events for each sow were analysed using continuous sampling. Data were analysed using a general linear model and were considered significant at $P<0.05$. There was no difference between SS pen treatments on sow behaviour at day 8 or weaning so pen data were pooled. Housing did not affect time taken to lie down ($P>0.05$). Sows from SS pens tended to initiate more interactions with piglets on day 8 ($P=0.08$) compared to farrowing crates. Farrowing crate sows decreased the frequency of nursing events as lactation progressed (8: 1.3 ± 0.1 versus wean: 1.0 ± 0.1 nursings/hour), whilst the nursing events in pens were similar across days (8: 1.3 ± 0.1 versus wean: 1.2 ± 0.1 nursings/hour) ($P=0.05$). This resulted in a tendency for farrowing crate sows to exhibit less nursing events at weaning ($P=0.07$). Reduced confinement in later lactation facilitates improvements in some aspects of maternal behaviour of sows. Future work should determine the impact of these changes in sow behaviour on piglet performance.

The effects of parity on the eating behavior and feed intake of milking cows kept in a free-stall barn

Shigeru Morita[1], Yuuki Yamane[1], Shinji Hoshiba[1], Ikuo Osaka[2] and Tamako Tanigawa[2]
[1]Rakuno Gakuen University, Ebetsu, 069-8501 Hokkaido, Japan, [2]Hokkaido Research Organization, Konsen Agricultural Experiment Station, Nakashibetsu, 086-1135 Hokkaido, Japan; smorita@rakuno.ac.jp

Cow management in a free-stall barn is generally based on free access to total mixed ration (TMR). Milking cows were divided into two groups according to the stage of lactation for matching the nutrient requirement in some farms. In the free access conditions, the eating behavior of cows differed with performance and parity. The objective of this study was to investigate the effects of parity on eating behavior and feed intake of cows in two stages of lactation. The cows were allocated into two groups within days after calving. In the early lactation period, cows were kept in one group and offered TMR (about 74% TDN and 16% CP on a DM basis). The cows in the later lactation period were given the TMR (about 69% TDN and 14% CP on a DM basis). Eating data was collected over an 86-day period using an automatic measuring system of the amount of intake and the behavior of eating for groups of rearing cows, individually. The meal criterion was calculated from the distribution of intervals between visits to trough. The average number of cows during the object period was 35 (11.6 primiparous cow, 14.3, 4.8 and 4.3 cows in 2nd, 3rd and after 4th parity) during early lactation and 30 (8.8 primiparous cow, 11.5, 7.1 and 2.6 cows, respectively) in the later lactation. The total quantity of data on eating was 3,012 cow-days in the early period and 2,583 cow-days in the later period. The parameters of eating behavior and the feed intake of each parity were compared using Kruskal-Wallis test. During the early lactation period, the average daily eating time by the primiparous cows was 280 minutes, and it was longer than that of the 3rd and after 4th parity cows. In the later lactation period, that of primiparous cows was 307 minutes, and it was longer than that of multiparous cows. The eating time of the 2nd parity cows was 288 and 281 minutes in the early and later lactation period. The eating time decreased with the number of parity in both stages of lactation. The amount of milk yield of the 2nd parity cows was highest during the early lactation period, and lowest in the later lactation period. In the later period, the feed intake for primiparous and 2nd parity cows were smaller than that of the 3rd and after 4th parity cows. The eating speed and meal size increased with the number of parity in both stages of lactation. The daily frequency of meals decreased with the number of parities in both lactation periods. The eating behavior and feed intake show similar changes with parity in both stages of lactation.

Effects of the short-term stress on hair cortisol concentrations in calves

Hideaki Hayashi[1], Chigusa Arai[1], Yurie Ikeuchi[1], Fukushi Subaru[1], Shigeru Morita[2] and Shinji Hoshiba[2]
[1]Rakuno Gakuen University, Department of Veterinary Physiology, School of Veterinary Medicine, 582 Bunkyodai Midorimachi, 069-8501, Japan, [2]Rakuno Gakuen University, Department of Sustainable Agriculture, College of Agriculture, Food and Environment Sciences, 582 Bunkyodai Midorimachi, 069-8501, Japan; hhayashi@rakuno.ac.jp

Dairy cattle suffer various stress in management and production, and it is necessary to reduce these stress as much as possible. Cortisol is a key hormone in the Hypothalamic-Pituitary-Adrenal axis (HPA) reaction, and blood and salivary cortisol are usually used for an index of the stress. However, blood and salivary cortisol exhibit circadian rhythm, and animals suffer stress in these sampling. Hair cortisol is not affected by the circadian rhythm, and reflects blood cortisol of the long term such as several weeks. Therefore in this study we aimed to investigate effects of the short-term stress on hair cortisol concentrations in calves. Twelve Holstein male calves aged 5 weeks were used in this study. After white and black hair were collected from the shoulder and the hip, calves were dehorned by electric calf dehorner. The hair was collected by cutting the root with scissors. Seven days later, after white and black hair were collected from the shoulder and the hip, calves were weaned. Furthermore, 21 days later, white and black hair were collected from the shoulder and the hip. Collected samples were extracted with methanol after washing with isopropanol, and measured with cortisol EIA kit. Statistical significance was determined using Paired t-test. Values were considered to be statistically significant if their P-value was <0.05. White hair cortisol concentration of shoulder increased with time, and the concentration of 7 days was significantly higher than 0 days ($P<0.05$). White hair cortisol concentration of hip also increased with time, and the concentration of 28 days was significantly higher than 0 days ($P<0.05$). In addition, white hair cortisol concentration of hip was significantly higher than the shoulder in 0 days ($P<0.05$). On the other hand, black hair cortisol concentration changed in neither shoulder and hip with time. However, black hair cortisol concentration of hip was significantly higher than the shoulder in 0, 7 and 28 days ($P<0.05$). From these, in this study, it is suggested that short-term stress affects white hair cortisol concentration, but does not affect black hair cortisol concentration. In addition, it is suggested that the level of hair cortisol build up varies according to a location. In conclusion, the results of this study indicate white hair cortisol concentration is available for short-term stress assessment such as acute stress.

Comparison of the stress level of cats kept at cat café's and at home using urinary cortisol

Ayuhiko Hanada, Ayaka Shimizu, Haruna Kubota, Natsumi Sega, Mika Kitazawa and Yoshie Kakuma
Teikyoukagaku University of Scince, 1-1 Minamiyuukarigaoka sakurasi, 2850859 Chiba, Japan; bell.dipp@gmail.com

The cat café is a style of business where multiple cats are kept loose and visitors pay to spend time with cats while having drink and food. The Japanese Act on Welfare and Management of Animals requires registration for this type of business as 'exhibiting of animals.' Recent amendment of the Act and the relevant Guideline set the regulation on exhibition time of animals although less strict for the cat cafés. There is a controversy if the café cats are stressed because of group living, being watched and touched by unfamiliar people regularly, and often staying in cages while the café is closed. Very little are reported on cat husbandry in this new business, so in this study, we collected urine of the cats and analysed their stress levels by measuring urinary cortisol as a physiological parameter and compared between café cats and pet cats. A questionnaire survey for café owners and cat owners was done to investigate their housing conditions for cats. The density of cats (head count per m^2 of living area), use of cage accommodation, the number and frequency of visitors were asked by questionnaire. The subjects were total of 75 cats, 51 from 9 cafés and 24 from 19 households. Urine was collected from each cat using non-absorbent litter and box with draining tray. The concentrations of urinary cortisol and creatinine were measured by EIA and a kit and urinary cortisol creatinine ratio (UCCR) was calculated. Comparisons of two groups were analysed by Mann-Whitney's U test. Spearman`s rank correlations were calculated to examine association between UCCR and the cat density, and the number of visitors. The cat densities at cafés were between 0.12-0.77 whereas 0.005-0.16 for households. The most common range of the number of visitors per day was 6-10 for cafés while 0-5 for all households. Forty-five percent of café cats were sometimes contained in cages, but only one cat at home was occasionally caged. The café cats were revealed living with higher cat density, more frequent and bigger number of visitors, and caged more often than those at home. The mean UCCR (X10^{-6}) was slightly higher for café cats (3.448) whereas 3.328 for pet cats, but they were not significantly different ($P=0.1679$). The UCCR was higher in cats that are not caged ($P=0.0323$) and also in café cats not using cages ($P=0.0008$), suggesting the environment not being disturbed by others may be less stressful for cats as reported by others. There were no significant correlations between UCCR and the cat density, and the number of visitors. Thus this study suggests that the café cats may not be more stressed than cats kept at home as family pets.

Communication about animal behaviour and training during routine veterinary appointments is reactive rather than proactive

Lauren C. Dawson, Cate E. Dewey, Elizabeth A. Stone, Michele T. Guerin and Lee Niel
Ontario Veterinary College, University of Guelph, 50 Stone Rd. E., Guelph, Ontario N1G 2W1,
Canada; dawsonl@uoguelph.ca

Veterinarians not only ensure the health of their patients, but also serve as a source of information for pet care and welfare. Behaviour problems and inappropriate training methods may have a significant impact on companion animal welfare, so a holistic approach to veterinary care should include regular inquiries about behaviour concerns. Our objective was to investigate the provision of behaviour and training advice by veterinarians during routine preventive ('wellness') appointments. A convenience sample of thirty veterinary clinics was visited. At each clinic, a veterinarian completed a written questionnaire and verbal interview. Video cameras were installed in exam rooms to capture veterinary appointments; up to six randomly selected appointments per clinic (81 appointments total) were analysed for discussion of relevant topics. Interviews revealed that although 93% of veterinarians feel confident in their ability to offer advice regarding behaviour and training, only 21% of participants discuss these topics proactively, whereas 66% discuss them only if a concern is presented or an issue arises during the appointment. Through the questionnaire, 63% of veterinarians indicated they routinely ask about behaviour issues during wellness appointments. While only 30% of veterinarians indicated they routinely discuss appropriate training methods, 93% claim to provide training advice to at least a subset of patients (i.e. dogs, young animals). Videos, however, revealed that behaviour problems were mentioned in only 20% of appointments. Advice regarding training was offered in 20% of appointments, including 23% of canine appointments and 33% of appointments with patients under two years of age. Results suggest that veterinarians recognize the importance of educating their clients and intend to do so, but in practice these conversations happen less often than they believe. Veterinarians also appear to take a reactive approach to discussions on behaviour, and thus, may miss the opportunity to prevent, identify and treat many behaviour issues.

Behavior of beef male calves after different castration methods in Uruguay

Marcia Del Campo and Juan Manuel Soares De Lima
National Institute of Agricultural Research, Ruta 5 Km 386, 45000, Uruguay;
mdelcampo@tb.inia.org.uy

This experiment was conducted to evaluate the stress response of beef calves after alternative methods to surgery without pain relief, the most used in this region. Thirty unweaned animals, 7 days old were assigned to three treatments: (1) handled/uncastrated (NC, n=10) or castrated by (2) surgery with local anesthesia (lidocaine, 8 ml; LID, n=10) or (3) rubber ring (RR, n=10). Behavior was observed by scan and focal sampling during 3 days after castration and every seven days, for 2 months. During Day 1, RR had a higher frequency of pain behavior (abnormal lying and standing) than LID. Is remarkable that in RR, they were mainly registered during the first 30 minutes after the procedure and no longer than 3 hours. The frequency of pain behavior was not different between LID and NC during Day 1. Moreover, 3 hours after castration there were no differences between Treatments ($P<0.05$, proc glimmix, SAS). During the second day, the frequency of abnormal lying and standing associated to head turning increased in LID, being higher than in RR ($P<0.05$, proc glimmix, SAS). Furthermore, RR decreased pain frequency during Day 2, being no different to NC and suggesting a shorter acute response when surgery is not involved. After 48 hours, there were no differences between castrated an NC calves. Is possible that the emotional bond with their mothers, partially contributed to a shorter/less intense stress response than those reported by other authors for intensive conditions. In conclusion, the use of anesthesia could have probably contributed to reduce the stress response after castration during day one, but pain was evident during Day 2. RR caused pain for half an hour after its application but it seemed to be a good alternative for 1 week old calves, since pain was not evident for more than 3 hours after the procedure.

Changes of plasma cortisol concentration, heart rate variability and behavior in Yonaguni pony by practical horse trekking

Akihiro Matsuura, Ayaka Ono, Tadatoshi Ogura and Koichi Hodate
Kitasato University, Animal Science, 23-35-1, Higashi, Towada, Aomori, 034-8628, Japan; matsuura@vmas.kitasato-u.ac.jp

Horse trekking (HT) involves taking a stroll on a horse along a walking trail in a forest, field and/or sandy beach. Usually in HT, horses exercise outside the riding facilities where they are kept. Therefore, they would be disturbed by the environment and the something strange on the trekking road. If they received such psychological loads, HT might become a critical issue from the viewpoint of animal welfare. In this study, the stress indices in horses were compared between the HT in shallow sea (S-HT), the HT in field (F-HT) and the resting state (Control; C). The plasma cortisol concentration, heart rate variability and behavior of seven Yonaguni ponies were analyzed. All procedure was carried out from 0900 to 1330 in consideration with their circadian rhythms. At S-HT, the ponies walked with the rider on their back in shallow sea (40 min) and walked/trotted without the rider between the stable and the seaside (30 min). At F-HT, they walked (60 min) and trotted (10 min) with the rider on their back along HT course. In each condition, three blood samples were taken before HT, immediately after and 2 hours after HT. Electrocardiogram and behavior were recorded before and after HT. At C, they were kept under resting state from 0900 to 1330, and same samplings were taken. Repeated measures ANOVA and Bonferroni multiple comparison tests were used for the statistical analyses. There was an interaction between exercise and time in cortisol ($P<0.05$), and the concentrations were decreased 2 hours after exercise compared with before exercise both in S-HT and F-HT ($P<0.05$). There was a difference in the LF/HF (index of sympathetic nervous activity), between exercises ($P<0.05$). The LF/HF tended to be lower in S-HT than in F-HT ($P<0.1$). The frequency of grooming behavior using forelimb was significantly lower in S-HT than in C ($P<0.05$). These results indicated that both S-HT and F-HT were not stressful tasks, especially S-HT caused little or no stress responses to the ponies.

Activity patterns and stress levels in domestic cats under different housing conditions:effects of light-dark cycle and space

Toshiki Maezawa, Ayaka Shimizu, Ayuhiko Hanada, Miyuu Kudo, Mika Kitazawa, Aya Yanagisawa and Yoshie Kakuma
Teikyo University of Science, Department of Animal Sciences, 2-2-1 Senjyusakuragi, Adachi-ku, Tokyo, 120-0045, Japan; g454002@st.ntu.ac.jp

Animal exhibitors such as pet shops are regulated by law and cats and dogs are now allowed to exhibit only from 8 am to 8 pm by recent revision of the Japanese Animal Welfare Act. As there is an exception measure for cats aged over 1 year, cat cafés are permitted to open until 10 pm. The influence of this practice is not clear because of shortage of previous research on the natural activity pattern of healthy domestic cats. Thus the purpose of this study was to examine activity patterns and stress levels of domestic cats when either light-dark cycle or space was changed. Six adult cats kept in a room at university went through all treatments. An accelerometer (HOBO Pendant G Logger, Onset Ltd., USA) was attached to the ventral location of the collar of each cat and acceleration was recorded every two seconds. Urine was collected non-invasively to measure cortisol and creatinine. Urinary cortisol/ creatinine ratio (UCCR) was calculated as a parameter of stress. Firstly, the influence of the change of the light-dark (LD) cycle was examined under two different conditions. Condition A was 12L12D (light period was 8 am to 8 pm) and Condition B was 14L10D (light period was 8 am to 10 pm), and each condition lasted for two weeks in turn. Activity was recorded by acceleration for 18 h/day except fixed care and feeding time in the morning and evening, and urine was collected on Day1, 4, 7, 10, and 14 for each condition. Change of space followed by freeing cats from cages at night (Condition C; outside cages from 8 pm to 8 am) or putting cats into cages at night (Condition D) while LD cycle was restored as Condition A. Each condition lasted for one week in turn. The activity level was recorded and urine was collected on Day1, 2, 3, 4, and 7 for each condition. As results, the mean±SE activity level (counts/day) for Condition A was 3036±485 and 3007±477 for Condition B, and there was no significant difference ($P=0.255$ d.f.=5 by t-test). The mean UCCR for Condition A was 2.92×10^{-6} and 2.96×10^{-6} for Condition B, and there was no significant difference ($P=0.402$). The mean±SE activity level (counts/day) for Condition C was 5138±808 and 3544±433 for Condition D and they were significantly different ($P=0.002$). The mean UCCR for Condition C was 3.67×10^{-6} and 3.71×10^{-6} for Condition D, and they were not significantly different ($P=0.242$). Thus activity levels of cats were not influenced by the change of light-dark cycle whereas activity was increased by freeing cats from cages at night. It was also suggested that these different housing conditions did not cause changes in the physiological stress response.

Effect of light color on behavior and physiological stress indicators in sheep

Risako Mizukami, Kazuna Nishikawa and Mizuna Ogino
Ishikawa Prefectural University, Nonoichi, 921-8836, Ishikawa, Japan; kimikoko365@gmail.com

Recent studies have shown that domestic animals could discriminate between different colors, and that light color affected their behavior and productivity. The advantage of green color stimuli was confirmed when cattle were conditioned with a food reward. Therefore, green light shows potential for domestic animal management. In this study, we examined the effect of different light intensities (high or low) and light color (white or green) on behavior and blood stress indicators in sheep. Furthermore, we used the same light intensity for each condition and investigated the effect of light color alone. In Experiment 1, we used 4 Suffolk sheep maintained in separate cages (1.1×0.6 m). The experiment was performed in September 2013. The following lighting conditions were used: no lighting (N: 0 lux), low lighting (L: 224±17.6 lux), high lighting (H: 4,206.1±441.1 lux), low green lighting (LG: 73.8±4.1 lux), and high green lighting (HG: 1,415.4±7.2 lux). In Experiment 2, we used 6 Suffolk sheep maintained in separate pens (1.3×1 m). The experiment was performed in August 2014. The following lighting conditions were used: white lighting (182±5.9 lux) and green lighting (166.1±7.3 lux). That sheep were able to show each other in experiment. Both experiments were performed from 21:00 to 03:00. Through the experiments, lighting were kept. Behaviors and posture of the sheep were recorded at 3-min intervals, and blood was collected from the jugular vein by using a catheter at 30-min intervals. Rhythm cycles of time series data were calculated using spectral analysis. In Experiment 1, rumination was observed many times and cortisol concentration and secretion were low in L and LG conditions. Abnormal behavior was observed in H condition, but not in HG condition. High light-intensity conditions may cause animals to express abnormal behavior, which could be suppressed by using green light. In Experiment 2, rumination was observed to a significant extent under green light ($P<0.05$). Furthermore, abnormal behavior observed under white light may be suppressed under green light. The rhythm cycle of plasma cortisol was 2 h, which was similar to that of plasma cortisol under a low-stress condition. Since higher rumination was observed under green light than under white light, a comfortable environment for ruminants could be provided by using green light, particularly under high-stress conditions.

Waddle this way: evaluating duck walking ability using a treadmill performance test

Christopher J. Byrd[1], Russell P. Main[2] and Maja M. Makagon[1]
[1]Purdue University, Department of Animal Sciences, 915 West State St., West Lafayette, IN 47907, USA, [2]Purdue University, Department of Basic Medical Sciences, 625 Harrison St., West Lafayette, IN 47907, USA; byrd17@purdue.edu

Gait scoring is a relatively inexpensive, easy-to-implement method commonly used for evaluating the occurrence and severity of walking impairments in poultry species. Despite their popularity, gait scoring systems are criticized for being subjective as a result of their reliance on descriptive definitions to rank walking ability. A large body of work has focused on comparing gait scoring systems against quantitative measures of walking abilities, particularly in chickens and turkeys. However, little attention has been given to systems used for evaluating ducks. We evaluated the relationship between gait score categories and the performance and behavior of commercial Pekin ducks using a treadmill test. Ducks from three gait score categories (GS0=smooth gait, n=55; GS0.5=labored walk without easily identifiable impediment, n=56; GS1=obvious impediment, n=59) were walked on a treadmill at a speed of 0.31 m/s. The amount of time each duck was able to walk at this pace was recorded. Three test termination criteria were used: (1) duck required support from the observer for more than 3 seconds in order to stay on the treadmill, (2) duck laid down on the treadmill with no effort to stand back up or evade the support of the observer's hand, or (3) duck reached the maximum trial time of 10 minutes. Data were analyzed using a general linear model in SAS. Tukey's multiple comparison test was used to assess treatment differences. GS1 ducks (lsmean=287.8 seconds) walked for significantly shorter amounts of time as compared with GS0.5 (lsmean=392.4 seconds; $P=0.004$) and GS0 (lsmean=475.3 seconds; $P<0.0001$) ducks. The average length of time that GS0 and GS0.5 ducks spent walking on the treadmill was also different ($P=0.032$). Video recorded during the test was evaluated for behavioral correlates (counts of sitting, stumbling, and leaning) hypothesized to be related to walking ability. The relationship between gait score and each behavior were analyzed using a negative binomial model for count data. No effects of gait score on the occurrence of these behaviors ($P>0.05$) were found. Overall, time spent walking was an effective measure for assessing quantitative differences in Pekin duck walking ability using a treadmill performance test. The treadmill performance test could be a valuable tool for assessing development of walking issues, the effectiveness of treatment, or as an additional tool for selection of leg health.

The relationship between Japanese stockperson's attitudes and productivity of dairy cows

Michiru Fukasawa[1], Masatoshi Kawahata[2], Yumi Higashiyama[1] and Tokushi Komatsu[1]
[1]NARO Tohoku Agricultural Research Center, Livestock and forage research devision, 4 Akahira, Shimo-kuriyagawa, Morioka, Iwate, 020-0198, Japan, [2]Aomori Prefectural Industrial Technology Research Center, Livestock Research Institute, Dairy and forage department, 51 Biwano, Noheji, Aomori, 039-3156, Japan; shakecat@affrc.go.jp

A stockperson has a unique role in dairy farming. Some studies have shown that a stockperson's positive attitude about cows and their jobs enhances dairy productivity. However, most of these studies were done in western societies. Attitudes about animals are considerably different between cultures. The aim of this study was to examine the relationship between stockpersons' attitudes about cows and dairy productivity in Japan. Three attitudes (positive, negative, and job satisfaction) were evaluated by an anonymous questionnaire that contained 24 items (9, 8, and 7 items, respectively). The stockperson who was responsible for farm management answered the questionnaire, and the responses for each item were rated by a five-point Likert scale that ranged from strongly agree to strongly disagree. The Likert scales were converted to numeric scores that ranged from 5 (strongly agree) to 1 (strongly disagree). Each attitude score was calculated by summing the numeric scores of the corresponding items. We distributed the questionnaire to 61 dairy farms that were participating in the milk-recording program in Aomori prefecture. Forty-four answers were received, but nine answers were removed because they were inappropriate. The following production data which was the average of test day during last 13 months before survey were collected from the milk-recording program; the number of milking cows in the herd, times of artificial insemination, milk yields, milk fat contents, protein contents, solid non-fat contents, milk urea nitrogen contents, concentration feed amounts, somatic cell counts, non-pregnant intervals, dry periods, and delivery intervals. Positive attitude scores correlated negatively with negative attitude scores and positively with job satisfaction scores, milk yields, and milk urea nitrogen contents. Negative attitude scores or job satisfaction scores did not correlate with any productivity traits. Three farmer groups were identified by a cluster analysis of the three attitude scores. Group P showed significantly higher positive attitude and job satisfaction scores than those of the other groups. Group N showed significantly higher negative and lower positive scores than those of the other groups. Group M showed moderate attitude scores. There were no differences in the ages of the stockpersons and the number of milking cows among the farm groups. The effect of farm group on milk yield tended to be significant, with the milk yield in Group P tending to be higher than that in Group N. These results suggested that stockpersons' positive attitudes for cows and their jobs improved dairy productivity in Japan.

Benchmarking dairy calf welfare and performance

Dax Atkinson, Christine Sumner, Rebecca Meagher, Marina A.G. Von Keyserlingk and Daniel M. Weary
University of British Columbia, Animal Welfare Program, 191 – 2357 Main Mall, V6T 1Z4 Vancouver BC, Canada; dax.atkinson@gmail.com

Poor health and growth of pre-weaned dairy calves can have lasting effects on development and production. This study aimed to benchmark calf rearing outcomes, report these findings back to producers, and document the effects of this intervention. Fifteen Holstein dairy farms were recruited. Blood samples were collected during the first week of life to estimate serum total protein (STP). Calf growth rates were estimated from biweekly heart-girth measurements. Following delivery of the benchmark report, a second period of data collection to assess the efficacy of management changes was conducted. Data collection is ongoing, but of calves tested to date (n=184), 12% had failure of passive transfer of immunity (defined as STP below 5.2 g/dl) with another 19% at sub-optimal STP levels (below 5.5 g/dl). It is recommended to feed a minimum of 4 litres of high quality colostrum within 6 hours to ensure successful passive transfer. On average (±SD), farms ensured a minimum of 3.1±0.8 litres of colostrum within a maximum of 10.6±3.3 hours. Fourteen farms had no protocol to assess the quality of colostrum. Repeat measurements of 247 pre-weaned calves (1-70 days) estimated a population growth rate of 0.65±0.33 kg/day. Calves provided sufficient milk have been shown to obtain growth rates of 1.0 kg/day. Mean daily milk provided was 7.1±2.2 l and ranged from 4 to 12.5 l, with 9 farms feeding below the recommended equivalent of 20% body weight. Additional results are pending; including how benchmarking influences producer decisions and calf outcomes, as well as an assessment of calf behavior across different management systems. Study participants expressed value in the benchmarking data and reported that the process stimulated reflection on current practices. These early results indicate a demand for practical on-farm assessment tools that lead to improved calf welfare.

Accurate sensing of pre-parturient cattle behaviour using IR depth-sensor camera

Haruka Sato[1], Daisuke Kohari[1], Tsuyoshi Okayama[1], Kasumi Matsuo[1], Takami Kosako[2], Tatsuhiko Goto[1] and Atsushi Toyoda[1]
[1]*Ibaraki University, College of Agriculture, Ami 4668-1, Inashiki-gun, Ibaraki, 300-0331, Japan,* [2]*NARO Institute of Livestock and Grassland Science (NILGS), 2 Ikenodai, Tsukuba, Ibaraki, 305-0901, Japan; 11a1025s@acs.ibaraki.ac.jp*

Dystocia and stillbirth strongly affect beef industry productivity. Prediction of parturition time in breeding cattle using some predictive instruments is helpful for breeding support. Although many sensors put on an animal's body can facilitate prediction of their parturition from changes of some pre-parturition behaviour, a more direct and non-invasive sensing method was formulated using IR depth-sensor cameras. This study clarified the sensing accuracy of IR depth sensor camera systems for assessing the pre-parturition behaviour of breeding cattle. Eight cattle delivered between May and November 2014 to the Field Science Center of Ibaraki University were used as subject animals. Two video cameras and an IR depth-sensor camera were placed at the calving pen. The authors observed four 24-h periods of pre-parturition behaviour such as posture change (standing, lying down), walking, circling, turning the head to the abdomen and tail lifting. The frequencies and percentages of these behaviours were calculated from the video observation data. Furthermore, these data from video and IR depth-sensor camera were compared and were used to calculate concordance rates. Data were analyzed using software (R ver. 3.1.2; The R Foundation for Statistical Computing). The posture change frequency and the percentage of walking were increased 6 h before calving. (The posture change frequency: 24-7 h 5.5-14.5 no./h, 6 h 18.5 no./h:Median. The percentage of walking: 24-7 h 0.06-1.39%/h, 6 h 1.43%/h: Median). Circling and turning the head to the abdomen were increased 3 h before calving (Circling: 24-4 h 0 no./h, 3 h 1.0 no./h: Median Turning: 24-4 h 5.5-18.5 no./h, 3 h 20.5 no./h: Median). Tail lifting was increased 5 h before calving(24-6 h 0.7-4.8%/h, 5 h 6.6%/h: Median). Posture, Circling and turning the head to the abdomen concordance rates were 80-100%. However, that of tail lifting decreased during 4 h before calving because their condition exhibited there were some differences. These results demonstrate that the IR depth-sensor camera had high accuracy for detection of pre-parturition behaviour of cattle. Detection of tail lifting requires more definition to discriminate their conditions.

Automatic monitoring system for ruminating activities of dairy cattle using a highly sensitive low-frequency pressure sensor

Akira Dokoshi[1], Tetsuya Seo[2] and Fumiro Kashiwamura[3]
[1]KONSEN Agricultural Experiment Station, 7 Asahigaoka Nakashibetsu Hokkaido, 086-1135, Japan, [2]Obihiro University of Agriculture and Veterinary Medicine, Inadatyou Obihiro Hokkaido, 080-8555, Japan, [3]Obihiro University of Agriculture and Veterinary Medicine, Inadatyou Obihiro Hokkaido, 080-8555, Japan; dokoshi-akira@hro.or.jp

The rumination behaviour of dairy cows could provide valuable information, thereby allowing the examination of the physical structure of the diet and the health status of dairy cows. This study was designed to develop a rumination monitoring device, which is wearable around the cow's neck, and to evaluate its accuracy for practical use in future. In experiment 1, we made a prototype of the rumination monitoring device with a highly sensitive pressure sensor for detecting low-frequency waves. The monitoring device was placed in ABS plastic case (size = 220×60×30 mm, weight=255 g), which contained a highly sensitive pressure sensor, wireless module and two rechargeable Ni-MH battery. A highly sensitive pressure sensor (SUD-571, SENTRANDGIKEN, Japan) can detect pressure of the frequency of 0.3~40 Hz. A 10-bit AD converter was used to convert sensor voltage output to integer value. Wireless module (transmission rate 4.8 kbps, available frequency bands 429 MHz) was used to send the integer value at regular intervals. The data are sent to the PC through a receiver. Date are processed and digitally stored within the PC. The distance that a device can transmit and receive is approximately 100 m. The device can be used continuously for about 30 hours before its batteries need to be replaced. The device placed in a leather bag for waterproofing and it was attached to the collar. In addition, counterweight (size=70×70×30 mm; weight=540 g) was attached to the collar to ensure the device retained its position about 5 cm behind the ear in the upper third neck in cows. The data obtained were categorized into ruminating, eating and non-chewing activities, followed by spectral analysis using the cepstrum technique. Characteristic peaks appeared in the frequency band around 1.3 and 2.6 Hz during rumination. We then developed a program for the automatic identification of rumination behaviour based on its characteristic. In experiment 2, behavioural data were collected from dairy cows in both free-stall and tie-stall barns, and rumination time measured by the device was compared with that obtained by visual observation to verify the accuracy and reliability of the device. The overall average accuracy rate of determination of rumination time was 91.1%, demonstrating that the device has the possibility of practical use.

Grooming desire of pre-weaned calves to stationary cow brush restriction

Yuki Kimura and Daisuke Kohari
Ibaraki university, college of agriculture, 4668-1, ami-machi, inashiki-gun, ibaraki, 3000331, Japan; 14am105n@vc.ibaraki.ac.jp

Cow brushes are used as enrichment devices to satisfy desires for grooming behaviour in cattle. Studies of such brushes have been conducted mainly with adult cattle, but only rarely with calves. Evaluating the desire for enrichment devices is necessary when using some devices. This study was conducted to clarify the desire for brush use by pre-weaned calves. At the Field Science Center, College of Agriculture, Ibaraki University, 10 calves of Japanese black cattle were used as subject animals (2-3 months, 7 male and 3 female calves). We used a stationary cow brush (B2; DeLaval) to which calves had been habituated for 2-4 weeks. Experiments were divided into three periods: control, removed, and replaced. Control and removed periods were set as seven days each. The replaced period was set as five days. During the control period, the calves were observed on the last day and one other voluntary day. During the removed period, the first day and voluntary one day and in replace period, we observed them on the first day and fifth day. We observed maintenance behaviour (eating, rest, rumination, etc.) and five grooming behaviours: self-grooming, grooming using some object other than the cow brush (object-grooming), social grooming, maternal grooming, and grooming using the cow brush (brush-grooming). We observed the calves for 6 h each day. Maintenance behaviour was recorded by 2 min scan sampling. Grooming behaviours were recorded using successive observation methods. Percentages of respective maintenance behaviours in each period were compared using chi-square tests. Durations of the self, object, social and maternal-grooming were compared using one-way repeated measures of ANOVA in each period. Durations of brush-grooming between the control and replaced period were compared using paired t-test. No difference was observed in maintenance behaviour, self, object and social-grooming in each period. However, maternal grooming tended to decrease during the replaced period (Control 77.4±67.4 s/calf/h; Removed 65.4±59.3 s/calf/h; Replaced 36.1±39.4 s/calf/h, $P=0.09$,one-way repeated measures of ANOVA). Brush-grooming was increased from the control to replaced period (2.6±3.3 s/calf/h, 20.6±14.4 s/calf/h, $P<0.01$,paired t-test). However, the sums of the durations of maternal and brush grooming were not significantly different. These results suggest that the desire of brush grooming is independent from self-, object-, and social grooming. Furthermore, maternal grooming and brush-grooming might share a complementary relation.

Effects of the resting-platform height and the ramp-slope angle on utilization by goats

Takeshi Yasue and Tomoyuki Wakai

Ibaraki University, School of Agriculture, Ami, Inashiki, Ibaraki, 300-0393, Japan; tyasue@mx.ibaraki.ac.jp

Since goats prefer to rest at elevated positions, developing a platform that would allow goats to rest higher off the ground would enable them to utilize their resting space more effectively. The present study examined the preferred height and slope angle of an adjustable platform that was installed in a goat feedlot. Fourteen Saanen goats (age: 0.4-6.9 years old, live weight: 14.0-53.9 kg) and 14 height and slope platform configurations were used in the study; platform height and slope was varied in increments of 10-30 cm and 5° respectively, in the range 60 cm/10° to 305 cm/75°. The ability of each individual to ascend/descend the slope was assessed using a food reward test in which goats were encouraged onto the platform using a reward (cereal feed). All goats were tested individually at one height and slope configuration on the same day, and voluntary use of the platform was observed using a video recorder during the resting phase (10:30-12:00 and 13:00-15:00) in the following another day. The 14 platform configurations were presented to the goats in three cycles: period 1 (13 Sep.– 30 Nov.), 2 (2 Dec. – 8 Jan.) and 3 (10 Jan. – 6 Feb.). The live weight and seven body measurements (body length, withers height, etc.) of individuals were determined in each period. The relationships between the maximum slope angle that each goat could ascend/descend, and their age, live weight, seven body measurements, and the relationship between the number of platform available goat and the voluntary platform use (total resting time on the platform / total observation time) at each configuration were examined by Spearman's rank correlation coefficient. The maximum slope/ height configuration that goats could ascend/descend in the food reward test ranged from 20°/115 cm for the relatively larger goats, to 70°/295 cm for the relatively younger and smaller goats. The maximum slope/height configuration of each goat did not vary among period 1 to 3. Significantly negative correlations were observed between the maximum slope angle and age, live weight, and the seven body measurements for each goat ($P<0.05$). Significantly positive correlation was observed between the number of platform available goat and the voluntary use of platform during the resting phase ($r_s=0.71$, $P<0.01$), and the voluntary use had become maximum (76% of observation time) in 20°/115 cm which was the steepest/ highest configuration that all goats could ascend/descend. It was concluded that the 20°/115 cm platform was the most desirable from the view point of utilization by grouped goats however the more steeper/higher platform was desirable in order to disperse vertically their spatial distribution according to their age and/or body size.

Long-term exercise on soil floor improves the health and welfare of Japanese Black steers

Sayuri Ariga[1,2], Shigehumi Tanaka[3], Takashi Chiba[3], Kyoichi Shibuya[3], Iori Kajiwara[2], Siyu Chen[2] and Shusuke Sato[1,2]

[1]Teikyo University of Sience, Department of Animal Science, 2-2-1, Senjusakuragi, Adachi, Tokyo, 120-0045, Japan, [2]Tohoku University, Laboratory of Land Ecology, Graduate School of Agricultural Science, 232-3, Yomogita, Naruko-Onsen, Osaki, Miyagi, 989-6711, Japan, [3]Tohoku University, Field Science Center, Graduate School of Agricultural Science, 232-3, Yomogita, Naruko-Onsen, Osaki, Miyagi, 989-6711, Japan; s_ariga220@yahoo.co.jp

We have previously reported that outdoor exercise on soil floor for 1-h promoted Japanese Black steers to perform normal behaviour for 2 days. In this study, we investigated the long-term health and welfare effects on steers of 1-h exercise on soil floor every 3 days. Steers in the early (E group: 15±2-month-old) and late (L group: 28±2-month-old) fattening phases were reared indoors on concrete floors covered with sawdust. Three E groups and three L groups with 3 steers per group were subjected to 3 treatments: without outdoor pen (WO), with outdoor pen with concrete floor (OC), and with outdoor pen with soil floor (OS). Blood samples were collected before each treatment (Day B), and at the first (Day F) and second (Day S) months after treatment was initiated. Metabolites and serum cortisol levels of samples were analysed. Carcass characteristics were recorded in L groups. Steers individually encountered to a novel calf decoy for 20-min, and their behaviours during this period were recorded. Metabolites and serum cortisol level were analysed by ANOVA. Carcass characteristics and behaviour were analysed by Kruskal-Wallis test. More steers had normal serum glucose concentration in L-OS treatment than in other treatments at Day S. Ten of 18 carcass characteristics were superior in OS than in other treatments (not significant). OS steers had lower basal cortisol concentration (WO: 1.5±0.8, OC: 1.8±0.6, OS: 0.5±0.3 μg/dl; $P<0.05$), and more cortisol response to tight restraining for 20-min than in other treatments at Day F (44±131, 48±60, 374±306%; $P<0.05$). The number of steers allo-grooming the decoy was greater in OS than in other treatments ($P<0.05$). The number of steers defecating during the novel decoy test was less in OS than in WO ($P=0.08$). In conclusion, OS steers were healthy and more affiliative to the novel object, and showed lower stress and fear level.

Effect of cow brush on the growth performance and behavior of the fattening Hanwoo steers

Kang Hoon Lee[1], Ka Young Yang[1], Gie Won Kim[1], Joo Hun Kim[1], Yeon Soo Park[2], Chang Woo Lee[2], Jae Jung Ha[3] and Young Han Song[1]
[1]*Kangwon National University, College of Animal Life Sciences, Animal Life Science Bldg 1-204, 200-701, Chun-cheon, Korea, South,* [2]*Gangwon-do Livestock Research Institute, 1028-3 Hyeancheon li, Doonae meon, 255-831, Hwangsung gun, Korea, South,* [3]*Gyeongsangbuk-do Livestock Research Institute, mook li, 750-871, Youngjoo-si, Korea, South; hirkdgns@kangwon.ac.kr*

This study was conducted to investigate the growth performance and lying behavior of fattening steers. Twenty-eight steers (7 heads/pen, aged 24-months-old) with an on average body weight of 619.8 kg in fattening period from a period of six months. Statistical analysis was carried out using the Mixed Procedure of SAS. For the control group, a cow brush was not implemented, while a cow brush was installed at the center of the right side of the pen for the treatment group. The average daily gain was measured in a two-month interval after the beginning of the experiment. In order to observe the lying, feeding, standing, walking, brushing, and mounting behaviors of the treatment group. The behavior of the Hanwoo steers was observed from the shoot images for 12 hours during the day (06:00 am – 6:00 pm) from the four installed CCTV cameras (IR LED Camera; APD-7070V), with the behavioral changes noted every two minutes. The behavioral characteristics of the cows were observed 2-4 in a month. Lying behavior was more evident in the treatment group than the control group during the entire experimental period ($P<0.01$). Feeding and standing were significantly higher in the treatment group than in the control group ($P<0.01$). The longer standing behavior in the treatment group than the control group was due to the longer feed time in the treatment group than the control group, which was proportionate to standing behavior. The walking and fighting behaviors were also significantly higher in the treatment group ($P<0.01$) than in the control group. From the above results, in the animal welfare type Hanwoo growing, a cow brush would give a great effect on the average daily gain from the fact. And the long feed and less unnecessary in the walking as well as less fighting frequencies in the treatment group were observed than in the control group.

Personality and environmental enrichment affect vocalisation in juvenile pigs

Mary Friel[1], Hansjoerg P Kunc[1], Kym Griffin[1], Lucy Asher[2] and Lisa M Collins[3]
[1]Queen's University Belfast, School of Biological Sciences, 97 Lisburn Road, Belfast, BT9 7BL, United Kingdom, [2]University of Nottingham, School of Veterinary Medicine and Science, Bonington Campus, Loughborough, Leicestershire, LE12 5RD, United Kingdom, [3]University of Lincoln, School of Life Sciences, College of Science, University of Lincoln, Riseholme Park, Lincoln, LN2 2LG, United Kingdom; mfriel06@qub.ac.uk

Vocalisation has been proposed as a promising measure of animal welfare as it can be non-invasively monitored and is known to reflect the affective state of the caller. However it is not known how individual characteristics, such as personality and gender, are related to vocalisation in many domestic species. Furthermore, studies investigating vocalisation have focused on acute welfare-impairing events, but it is unknown how long-term contexts, such as living in a barren environment, affect acoustic signalling. This study investigated the relationship between these factors and vocalisation in the domestic pig. A total of 72 commercial crossbreed 6 week old pigs were tested in 4 replicates. Half the subjects were housed in barren pens and the other half were housed in enriched pens. Personality was assessed in repeated social isolation tests and repeated novel object tests; individual vocalisation was also recorded in these tests. Several behaviours were found to be repeatable and highly correlated; these were aggregated to create the continuous coping style scale. A linear mixed effects model was used to investigate the fixed effects of personality, gender and environment on vocalisation rate. Vocalisation rate was found to be significantly predicted by personality, with more proactive individuals signalling at a higher rate than reactive individuals ($P<0.000$). There was a significant interaction effect of environment and gender on acoustic signalling rate ($P<0.05$), with males from the barren environment having a significantly lower signalling rate than females from both environments and males from the enriched environment. These results indicate that personality should be taken into account when using vocalisation rate as a welfare measure. The fact that environment had an effect on vocalisation rate in males also indicates that acoustic signalling may be a sensitive measure of long term welfare status induced by environmental quality.

Behavior difference of giant anteaters (*Myrmecophaga tridactyla*) kept under the same captive condition

Yu Nakayama[1], Daisuke Kohari[1,2], Takashi Kimura[3], Eri Okuyama[3] and Michiko Kasahara[3]
[1]Tokyo University of Agriculture and Technology, 3-5-8 Saiwai-cho, Fuchu-shi, Tokyo 183-8509, Japan, [2]Ibaraki University, 4668-1, Ami, Ami-machi, Inashiki-gun, Ibaraki 300-0331, Japan, [3]Ueno Zoological Gardens, 9-83, Ueno Park, Taito-ku, Tokyo 110-8711, Japan; elenx2c@hotmail.co.jp

The stereotypic behavior of zoo animals frequently occurs due to the restricted display environment different from their habitat. However, the relationship between the enclosure condition of animals and their stereotypic behavior is not clear. This study aims to find out how the same enclosure condition affects two different giant anteaters' behaviors by comparing the appearances of their maintenance and stereotypic behaviors. We observed one male giant anteater (A) at Ueno zoo in 2014 and recorded the percentages of maintenance behaviors and the rate of stereotypic pacing. Furthermore, we recorded A's behaviors by three video cameras set in the enclosure and analysed by the 2D DLT Method of video analysing system for the purpose of measuring the movement locus, moving distance and movement speed. The data of A were compared with those of the female anteater (B) researched by Kohari et al. in same enclosure. The maintenance behaviors were significantly different between the two anteaters (X^2=80.5, $P<0.01$). Sleeping was observed only with A. The percentages of exploring and resting were higher with A than with B. The pacing rate of A was lower than that of B. The moving distance was less with A than with B (A: 3,789 m/6 h, B: 8,544 m/6 h). The movement speed was slower with A than with B (A: 0.23 m/s, B: 0.40 m/s), but their moving speeds both turned slower after feeding. One of the distinctively frequent walking pattern appeared in the feeding area for 30-60 minutes before the feeding time. This study suggests that the maintenance behaviors and the stereotypic pacing of giant anteaters kept under the same enclosure condition are significantly different, while there is similarity in their behavior before the feeding time.

Effect of a landscape immersion exhibit on the behaviour of captive Japanese macaques (*Macaca fuscata*)

Azusa Yatsushiro[1], Shiori Watanabe[2], Izuru Yoshimura[2], Misaki Furuie[2], Yasutaka Motomura[2], Shuho Hori[1], Atsushi Matsumoto[3], Masayuki Tanaka[4] and Shuichi Ito[2]
[1]Tokai University, Graduate school of Agriculture, Kawayo, Minamiaso-mura, Aso-gun, Kumamoto-ken, 869-1404, Japan, [2]Tokai University, School of Agriculture, Kawayo, Minamiaso-mura, Aso-gun, Kumamoto-ken, 869-1404, Japan, [3]Kumamoto City Zoological and Botanical Gardens, 5-14-2, Kengun, Higashi-ku, Kumamoto-shi, Kumamoto-ken, 869-0911, Japan, [4]Kyoto City Zoo, Okazaki Koen, Okazaki HOushojicho, Sakyo-ku, Kyoto-shi, Kyoto-fu, 606-8333, Japan; 9anz2105@gmail.com

Recently, zoo exhibits are changing from a traditional wire cage style to a naturalistic style. Landscape immersion, a naturalistic style of animal exhibit, is intended to give visitors an impression of actually being in the animals' habitat. At the Kumamoto City Zoological and Botanical Gardens, a Japanese macaque exhibit was changed from the old cage-style to a landscape immersion exhibit in October 2013. Landscape immersion exhibit has many plants and natural objects, we hypothesized that the macaques behaviour and welfare would be affected to Landscape immersion exhibit. This study investigated the behaviour of Japanese macaques. A total of 11-14 macaques were housed in the zoo. Five individuals were chosen for observation. The macaques were observed in the old cage-style exhibit from August to September 2013. The cage-style exhibit had a concrete floor and wooden tower. Subsequent to the re-housing to the landscape immersion exhibit, we observed the animals at one week, two-nine weeks, six months and a year. The macaques were observed using the scan sampling method at 3-min intervals from 09:00 to 17:00. The data of all individuals for each category were compared using Dunn's multiple comparison tests. After one week, the proportion of time spent on feeding was significantly less ($P<0.01$) in the landscape immersion (10.0±3.8%) than that in the cage-style exhibit (25.9±5.4%). The movement of the macaques was significantly greater ($P<0.001$) in the landscape immersion after a week (47.9±13.5%) than that in the cage-style exhibit (17.2±6.8%). Self-grooming and affiliative behaviour were significantly less ($P<0.05$ and $P<0.01$, respectively) in the landscape immersion after a week (2.0±2.6% and 0.3±0.4%, respectively) than in the cage-style exhibit (9.6±3.7% and 16.7±6.0%, respectively). Agonistic behaviour was significantly less ($P<0.05$) in the landscape immersion exhibit after six months (0.3±0.4%) than that after two weeks (1.1±0.5%). These results suggest that the re-housing of the captive Japanese macaques to the landscape immersion exhibits had some effects on their behaviour. However, their behaviour reverted to the cage-style behaviour after two weeks.

The role of uropygial gland in sex behavior and the morphology of the lateral nasal gland duct in domestic chicken

Atsushi Hirao[1], Masato Aoyama[2], Shu Takigami[3], Naoki Tsukahara[4] and Shoei Sugita[2]
[1]Jichi Medical University, School of Nursing, 3311-159 Yakushiji, Shimotsuke city, 329-0498 Tochigi, Japan, [2]Utsunomiya University, Department of Animal Science, 350 Minemachi, Utsunomiya city, 321-8505 Tochigi, Japan, [3]Kyorin University, Department of Health Science, 476 Miyashota-cho, Hachioji, 192-8508 Tokyo, Japan, [4]SOKENDAI, The Center for the Promotion of Integrated Science, Shonan Village, Hayama, Kanagawa, Japan; jhirao@jichi.ac.jp

Vomeronasal organs are not observed in birds. However, domestic chickens Gallus domesticus have a paired tubular organ at the base of the nasal septum called the lateral nasal gland duct (LNGD), which is anatomically similar to vomeronasal organ found in other vertebrates. This raises the question of whether domestic chickens can detect and utilize chemosignals from the uropygial gland. This study therefore aimed to clarify both the role of the uropygial gland on chicken sex behavior and the morphological features of the LNGD. In Experiment 1, we observed 10 intact male chicken's sex responses to female chickens treated with uropygial glandectomy (UGX) or sham operation (UGS) for 4 h trials. Wilcoxon signed-rank test to compare the number of each male's sexual responses to UGS and UGX for each 4 h trials, and we found that mounts and copulations of intact male birds were significantly more frequent with UGS than UGX ($P<0.05$ and $P<0.02$ respectively). In Experiment 2, male birds were treated with olfactory bulbectomy (OBX; n=9) or sham operation (OBS; n=12), and randomly selected individual birds were placed in a pen containing UGX and UGS birds. Observational data was collected as in Experiment 1 and Friedman two-way analysis of variance by ranks was used to determine the relationship between frequency of male sexual responses and male olfactory bulbectomy or preference for specific females. Mounts and copulations of OBS birds were significantly more frequent with UGS birds than with UGX birds ($P<0.02$, and $P<0.01$, respectively), but OBX males showed no difference in preference between UGS and UGX females. In Experiment 3, morphological study was conducted on the LNGD of 13 domestic chickens. Immunohistochemical staining revealed that several epithelial cell types expressed the G protein α subunit Gα olf/s, and that some cells expressed trace amine associated receptor 2 (TAAR2). The results of the three experiments show that mating choice in domestic chicken is affected by the female uropygial gland and male olfaction, as indicated by male sex behavior and the detection of Gα olf/s and TAAR2 in the LNGD. The female's uropygial gland therefore may act in chemosignaling in domestic chickens.

Choice behavior on feeding *Eucalyptus* leaves in koalas *Phascolarctos cinereus*

Tadatoshi Ogura[1], Sayaka Otani[1], Ayumi Takeishi[1], Riko Tateishi[1], Motko Ohata[1], Keizo Arihara[1], Tetsuo Nakayama[2], Keiko Yamabe[2], Hiromi Shigeno[2], Yoshiaki Tani[2] and Akihiro Matsuura[1]
[1]Kitasato University, School of Veterinary Medicine, 23-35-1, Higashi, Towada, Aomori, 034-8628, Japan, [2]Higashiyama Zoo and Botanical Gardens, 3-70, Higashiyama-motomachi, Chikusa-ku, Nagoya, 464-0804, Japan; togura@vmas.kitasato-u.ac.jp

Koala is endangered in some parts of the wild habitats, although it is an attractive species for public. They feed *Eucalyptus* leaves, which contain low nutrient content and high toxic components, in captivity as well as in the wild. In the genus *Eucalyptus*, approximately 600 species, which vary in the shape, size, hardness, odor of these leaves are present. Based on long husbandry experience, zoo keepers consider that koalas have preference for *Eucalyptus* species and feed specific species selectively. The factors involved in feeding choice and the cues for such choice are, however, not well understood. In order to approach them, this study investigated feeding behavior and feeding record in koalas housed in Higashiyama Zoo and Botanical Gardens. We observed 5 koalas for 8 days per subject from August 13 to 28 in 2014 from the public area. During this period, they were fed 3 species of *Eucalyptus* per day. The observation was conducted for an hour just since feeding started. The duration of feeding behavior and the *Eucalyptus* species which koalas fed were recorded using an instantaneous sampling method with a 30-second interval. During observation, the choice behavior when koalas fed leaves at the first time and started feeding again after leaving from leaves was video recorded. We also analyzed feeding records documented the fed species, place of cultivation, age of tree, the number of the day since the leaves arrived at the zoo, the weight of the fed leaves and the leftovers to examine the factors which affect the koalas' preference from January to December 2013 by the zoo keepers. The subjects showed significant difference in the Eucalyptus species ($\chi^2=491.1$, $P<0.001$) regarding the duration of feeding behavior. *Eucalyptus punctata* was one of the preferred species among 10 fed species. When the subjects chose leaves, behavior of 'smell and eat' was observed more often than that of 'eat without smelling' ($\chi^2=23.16$, $P<0.001$). Behavior of 'smell and leave' was also observed, suggesting that they discriminated the preferred leaves by odor cues. The analysis of feeding records showed that *Eucalyptus* species affected the preference for leaves ($\chi^2=37,565$, $P<0.001$). The number of days since the leaves arrived at the zoo also influenced the preference ($\chi^2=4167$, $P<0.001$). These results indicated that captive koalas have a preference for *Eucalyptus* species and the number of the days since the leaves arrived at the zoo and discriminate them based on smell.

Two-level cages promote the expression of natural behaviour in nursing rat dams

Wendy Ellison[1], Melissa Liu[1] and Sylvie Cloutier[2]
[1]*Washington State University, Center for the Study of Animal Well-being, P.O. Box 647620, Pullman WA99164-7620, USA,* [2]*Canadian Council on Animal Care, 800-190 O'Connor Street, K2P 2R3, Ottawa, Ontario, Canada; wendy.ellison@email.wsu.edu*

In their natural environment, nursing rats spend time away from the nest and the pups. In a standard laboratory cage the performance of this behaviour is limited by the size and layout of the cage. We hypothesized that rat dams, provided access to an area not easily accessible to the pups, would spend more time away from their pups as they progress towards weaning, thus expressing more natural behaviour. We assessed the response of Long-Evans rat dams (n=16) housed with their litters in either a standard cage (equipped with a small loft, hanging from one side of the cage, easily accessible to the pups, L1) or a two-level cage (with a shelf not easily accessible to the pups, L2). Dam activity (active vs inactive), pup-oriented behaviours (nursing and grooming) and location (shelf/loft vs floor) were recorded every 5 min for 24 h on days 5, 10, 15 and 20 after birth, and at weaning (day 21). Dams were the unit of analysis. Generalized linear mixed model ANOVAs were used. Activity was affected by cage type ($P=0.002$), L2 dams being less active than L1 dams (L1: 22.3±0.87% of scan; L2: 19.2±1.10). Activity decreased with time for all dams ($P=0.004$, Day 5: 22±1.7; Day 10: 23±1.8; Day 15: 19±0.8, Day 20: 16±1.4). Nursing and grooming were not affected by cage type ($P>0.05$) but varied with time ($P≤0.0001$). Nursing was more frequent on day 20 than 5 and 10 (Day 5: 2.2±1.12; Day 10: 3.5±1.28; Day 15: 6.4±0.90, Day 20: 9.3±1.83), whereas grooming was more frequent on day 20 than 5 and day 10 than 15 and 20 (Day 5: 0.65±0.309; Day 10: 0.30±0.187; Day 15: 1.55±0.270, Day 20: 1.59±0.241). Time on the upper level (loft or shelf) was affected by cage type ($P=0.03$), time ($P<0.0001$) and their interaction ($P=0.0006$). Overall, L2 dams spent more time on the upper level than L1 dams. Time spent on the upper level increased by approximately 13 and 32%, for L1 and L2 dams, respectively, as pups aged but decreased at weaning, especially for L2 dams (mean±SE, Day 5: L1: 8.2±2.36, L2: 9.9±3.0; Day 10: L1: 10.6±2.52, L2: 11.3±4.33; Day 15: L1: 21.3±3.79, L2: 32.2±4.74; Day 20: L1: 21.5±3.66, L2: 42.3±5.06; Day 21: L1: 14.6±2.54, L2: 17.3±3.07). These findings show that providing a cage environment with an area not easily accessible to the pups allows rat dams the opportunity to spend time away from the pups, most likely resting, as done in their natural environment. Our results have implications for the design of breeder cage environment by showing the need for providing dams with a resting space away from the pups.

Effects of environmental enrichment and zoo visitors on the stereotyped behaviour of a polar bear

Kosuke Momita[1], Atsushi Matsumoto[2], Hiroshi Anami[2] and Shuichi Ito[1]
[1]Tokai University, School of Agriculture, 869-1404, Kawayo Minamiaso-mura, Aso-gun, Kumamoto-ken, Japan, [2]Kumamoto City Zoological and Botanical Gardens, 869-0911, 5-14-2, Kengun, Higashi-ku, Kumamoto-shi, Kumamoto-ken, Japan; kskmomi@gmail.com

Polar bears (*Ursus maritimus*) are known to develop repetitive pacing behaviours (RPB), usually termed as stereotypic in a zoo environment. We investigated the effect of environmental enrichment and the presence of zoo visitors on RPB of a polar bear in the Kumamoto City Zoological and Botanical Gardens in Japan. Observation sessions were conducted for 10 days (6 days were open and 4 days were closed for visitors). We created three conditions using a combination of enrichment toys and feeders. These enrichment devices had installed two or four times a week randomly from two months before the observation. In condition 1, four enrichment toys were used. In condition 2, in addition to these toys, one input type enrichment device and two types of foraging enrichment devices (foraging balls and sliced apples on a wall) were used. In condition 3, the enrichment devices were same as condition 2 except that the amount of food for two foraging enrichment devices was doubled. For open days, we conducted the observation for 2 days for each condition. For closing days, we observed for two days for condition 1 and 3 each. Although we cannot explain the reason, RPB was rarely observed in one day of condition 2 of open day. We excluded this data from the analysis. On open days, the proportion of time spent on RPB was 48±4.7% in condition 1, 55% in condition 2, and 35±2.0% in conditions 3. On closing days, those were 35±3.2% in condition 1 and 17±2.8% in condition 3. The total durations of RPB per 25 min were not different between those before (5.8±4.2 min) and after (6.0±4.5 min) the introduction of input type enrichment device in condition 2 and 3 together. On the other hand, the total durations of RPB per 25 min before the introduction of foraging enrichment devices (7.3 min in condition 2 and 9.4±9.1 min in condition 3) decreased to 0 min after the introduction. The reduced RPB, however, increased soon afterward in condition 2. These results suggest that the foraging enrichment could effectively reduce RPB when the amount of food given is enough. In addition, zoo visitors may have a negative impact on the welfare of polar bears.

Dog's response to the 'still-face' interaction with human

Akitsugu Konno, Makoto Takeuchi, Masaaki Tezuna, Haruka Iesako, Ai Hamada and Shinji Yabuta
Teikyo University of Sciences, Department of Animal Sciences, 2525 Yatsusawa, Uenohara, Yamanashi, 409-0193, Japan; akitsugukonno@ntu.ac.jp

The 'still-face' paradigm has been used for studying human infant's response to communicative interaction with an adult. Infants generally avoid making eye-contact and show a reduced expression of positive emotion when exposed to the still-face and non-reactive adult, which is known as the still-face effect. To examine dog's responsiveness to human social signals, we tested how dogs reacted to the still-face interaction performed by a human partner. A total of 22 dog-owner pairs participated in this study. Dog's behavior was monitored in a three-step interactive episodes with a person: (1) the baseline episode in which the person 'played' with the dog in a normal way, (2) the 'still-face' episode in which the person looked at the dog with a neutral facial expression but never responded to the dog, and (3) the reunion episode in which the person resumed a normal interaction with the dog. We also tested the effect of familiarity of the human partner (i.e. owner vs non-owner) on dog's behavior. We predicted that the dogs would change their communicative behaviors (i.e. eye-contact and physical contact with a person) according to the different episodes, and that dogs would be more sensitive to the interactions performed by an owner than those by non-owner. A two-way ANOVA for repeated measures indicated a significant effect of episode on total duration of dog's eye-contact with a person (F $(2, 42)$=23.2, $P<0.01$), total duration of dog's physical contact with a person ((F $(2, 42)$=79.0, $P<0.01$)), and total time of dog's sniffing at ground ((F $(2, 42)$=19.0, $P<0.01$)), respectively. Post-hoc analyses indicated that dogs made eye-contact and physical-contact with a person for shorter periods of time and sniffed at ground for longer periods of time during the 'still-face' episode than during the baseline and the reunion episode. A significant effect of familiarity of human partner was found only in total duration of dog's eye-contact with a person ($P<0.01$), with dogs looking at an owner for longer periods of time than they looking at a non-owner. There was no significant effect of interaction in any of three behavioral measures. The still-face effect was also found in dogs, regardless of familiarity of human partner, indicating that dogs have a capacity of modifying their behavior depending on general communicative signals given by human partners. The still-face paradigm is useful for assessing dog's sensitivity to cross-species interactions. Future study may reveal relationships between individual differences in dog's communicative ability and previous experiences of socialization with humans.

Latency of pet rabbits to come out from their own cages at home or carries at novel place

Asami Tsuchida, Narumi Akiba, Mari Morimoto and Koji Masuda
Tokyo University of Agriculture, 1737 Funako, Atsugi-shi, Kanagawa, 243-0034, Japan; a3tsuchi@nodai.ac.jp

To clarify the factors about pet rabbits (Oryctolagus cuniculus) to give them some influences, 4 conditional experiments were performed with 8 pet rabbits (3 males, one was neutered, and 5 females, 9-month-old to 9-year-old, 4 rabbits were reared in solitary) kept by 6 owners. The experiments were carried out at their own home or a novel place with owners or strangers. We recorded with video-camera and measured a latency to come out from their own cages at home or carries at the novel place with owners or strangers sat away 1 m from the front of cages or carries. Every conditioned experiment was performed among 4-6 pm and the video-camera was operated by the owner in owner-condition and by an experimenter in stranger-condition. When a rabbit didn't come out over 30 min. after opening the door of cage or carry, the experiment was finished. One male (9-year-old) wasn't tested in novel-condition, because its physical condition was considered. In the results, 2 males and 2 females didn't come out in novel-condition, and in home-stranger-condition 1 female didn't come out. The latency in home-condition was significantly shorter than that in novel-condition (Mann-Whitney's U test, $P=0.003$). There was no significant difference in the latency between with owner and with stranger ($P=0.725$). In home-condition, the latency of males was significantly longer than that of females ($P=0.01$). Moreover, in home-stranger-condition, the latency of males was significantly longer than that of females ($P=0.034$), in compared with that there was no significant difference in the latency between males and females in home-owner-condition ($P=0.180$). The latency of rabbits unexperienced taking a walk (2 males, 1 female) was significantly shorter than that of those experienced taking a walk (2 males, 1 female)($P=0.014$). However, this result might be influenced by sample numbers of each sex. Rearing condition didn't show any significant differences on the latency.

The number of visitors and noise level affect behaviours of red panda in a zoo

Ken-ichi Takeda[1], Yukari Ishikawa[2] and Tajima Mitsuru[3]
[1]Institute of Agriculture, Academic Assembly, Shinshu University, 8304 Minamiminowa, Nagano, 399-4598, Japan, [2]Faculty of Agriculture, Shinshu University, 8304 Minamiminowa, Nagano, 399-4598, Japan, [3]Nagano Chausuyama Zoo, 570-1, Shinonoi, Nagano, 388-8016, Japan; ktakeda@shinshu-u.ac.jp

Many zoos exhibit animals in glass-walled enclosures, enabling visitors to see the animals up close. However, this might increase stress for the animals, as people can be noisy. Therefore, this study examined the behaviour of zoo-housed red pandas to establish whether visitor numbers and noise affected the animals' welfare. Eight red pandas (*Ailurus fulgens fulgens*) kept in the Chausuyama Zoo, Nagano, Japan, were used our study. During an opening the zoo, three red pandas were exhibited in indoors enclosure and five red pandas were in outdoors, respectively. The visitors cloud watch them both enclosure. They were observed for 4 days each on weekdays and holidays (total 16 days). In each enclosure, there was a place where the red panda is hidden and a tall tree. Focal animals were observed by the naked eye, using scan sampling methods, while they were on exhibit from 09:30 to 16:30. The behaviours and positions of the focal animals in the enclosure and the number of visitors in the viewing area were recorded at 5-minute intervals. The noise level and temperature-humidity index (THI) around the enclosure were recorded continuously. The vigilance of the pandas was positively correlated with the magnitude of the noise (rs=0.18, $P<0.05$) and the number of visitors (rs=0.12, $P<0.05$). The magnitude of the noise was also positively correlated with the frequencies of stereotyped behaviour (rs=0.20, $P<0.05$) and racing around madly (rs=0.13, $P<0.05$). When there were more visitors, the red pandas tended to remain in the rear of the enclosure rather than within 1 m of the observation window. These results indicate that captive red panda exhibited in glass-walled enclosures are under psychological stress. Visitors need to be quiet and the animals should have places to hide in the enclosure.

Eating behavior in thoroughbreds fed soaked beet pulp

Yousuke Matsuya[1], Yuuka Takisawa[2], Yukari Yamanaka[2], Harutaka Murase[3], Masataka Tominari[3] and Masahito Kawai[2]

[1]*Hokuchiku Co., Ltd., 328-66, Nishihorobetsu, Urakawa-Cho, 057-0002, Hokkaido, Japan,* [2]*Obihiro University of Agriculture and Veterinary Medicine, 2-11, Nishi, Inada-Cho, Obihiro-Shi, 080-8555, Hokkaido, Japan,* [3]*Japan Racing Association, Hidaka Training and Research Center, 535-13, Nishicha, Urakawa-Cho, 057-0171, Hokkaido, Japan; matsuya@hokuchiku.com*

To investigate the eating behavior in thoroughbred fed soaked beet pulp, intake and eating time were measured. Four thoroughbreds were fed whole oats (WO) or beet pulp (BP) each 4 kg/day, which was soaked in water on 2 times the amount of volume (Experiment 1). Four thoroughbreds ware fed rolled oats (RO) or beet pulp each 4 kg/day, which was soaked on 3 times (BP3) or 20 times (BP20) the amount of mass (Experiment 2). Beet pulps were soaked in water for 12 hours and BP20 was fed after cutting the water through a 2 mm sieve. Timothy hays were offered ad libitum, and each experiment had a control group (HAY1, HAY2) without feeding oats or beet pulp. Horses were kept in each stall and sandy paddock for 6 hours. The data were analyzed by ANOVA and Fisher's LSD. (Experiment 1) The DMI of concentrates by WO and BP were both 3.6 kg/day, and the voluntary DMI of hay were less than that by HAY1 (6.7 and 5.5 vs 9.4 kg/day, $P<0.05$). The eating time of concentrates by BP was longer than that by WO (115 vs 63 min/day, $P<0.05$), while that of hay by WO and BP were shorter than that by HAY1 (473 and 469 vs 636 min/day, $P<0.05$). (Experiment 2) The DMI of concentrates by BP20 was less than that by RO and BP3 (2.2 vs 3.2 and 3.1 kg/day, $P<0.05$). The voluntary DMI of hay by BP20 tended to be more than that by RO and BP3 (7.3 vs 6.6 and 6.4 kg/day), while these were less than that by HAY2 (10.2 kg/day, $P<0.05$). The eating time of concentrates by BP3 and BP20 were longer than that by RO (97 and 71 vs 38 min/day, $P<0.05$). The investigation indicates that feeding beet pulp prolong to the eating time however the amount of the added water considered to be affected the voluntary intake and eating behavior.

The maternal stress effects of an omega 3 enriched diet in chickens

Elske N. De Haas[1,2], Ludovic Caladreau[1,2], Elisabeth Baéza[3], Pascal Chartrin[3], Rupert Palme[4], Anne-Sophie Darmaillacq[5], Ludovic Dickel[5], Sophie Lumineau[6], Cécilia Houdelier[6], Cécile Arnould[1,2], Maryse Meurisse[1,2] and Aline Bertin[1,2]
[1]*Université François Rabelais de Tours, 60 rue du Plat D'Etain, 37000 Tours, France,* [2]*INRA Val de Loire, UMR85 Physiologie de la Reproduction et des Comportements; CNRS-UMR 7247; IFCE, 37380, Nouzilly, France,* [3]*UR83 Unité de Recherches Avicoles, 37380, Nouzilly, France,* [4]*Vetmeduni, 1210, Vienna, Austria,* [5]*Université de Caen Basse-Normandie, Groupe Mémoire et Plasticité comportementale, 14032, Caen, France,* [6]*Université de Rennes 1, UMR CNRS 6552, Ethos, 35042, Rennes, France; endehaas@gmail.com*

The n-3 and n-6 fatty acid (FA) content of the maternal diet during the prenatal period influences brain and behavioural development in mammalian offspring. Although, maternal effects on offspring behaviour are well characterised in birds, maternal dietary effects of n-3 and n-6 FA still deserve consideration. This study examined effects of maternal diet with high levels of n-3 or standard n-3/n-6 FA levels (i.e. control (C)) on the FA composition and hormone concentrations in yolk, maternal basal faecal corticosterone metabolites (FCM), and offspring's development and behavioural responses to novelty (novel food, novel object and novel environment), in white leghorn laying hens (n=18/group, offspring, n=48/n-3 vs n=50/C). The n-3 enriched diet increased yolk progesterone ($1,145\pm151$ vs 669 ± 95 ng/g, $P=0.02$) and estradiol concentrations (10.5 ± 0.5 vs 7.4 ± 0.3 ng/g, $P=0.002$), and n-3 FA content in yolk while reducing essential arachidonic acid compared to control diet (0.6 ± 0.05 vs 1.8 ± 0.2% of total FA, $P<0.01$). The hens fed the n-3 FA enriched diet had higher FCM levels than hens fed control diet (397.5 ± 56.8 vs 98.7 ± 19.1 ng/g, $P=0.003$). Chicks obtained from hens fed the n-3 enriched diet (n-3 chicks) weighted less at hatch than chicks obtained from hens fed the C diet (C chicks): 39.4 ± 3.4 vs 40.8 ± 3.3 g ($P=0.02$). N-3 chicks spent less time eating novel foods, especially females (12 ± 4 vs 23 ± 5 s, $P=0.01$,), and took longer to approach a novel object (173.9 ± 6 vs 144.6 ± 11.56 s, $P=0.01$). In a novel environment, n-3 chicks walked less than C chicks (6.0 ± 1.7 vs 11.4 ± 2.3 line crosses, $P=0.02$). These data suggest a stress effect of the maternal n-3 enriched diet, by increasing steroid yolk-hormones levels and FCM. They highlight a potential trans-generational detrimental impact of n-3 enriched maternal diet on offspring emotional reactivity – a potential new pathway for maternal effects in birds via the n3/n6 ratio in the maternal diet. ANR project PReSTO'Cog (ANR-13-BSV7-0002-02).

Unhusked rice feeding and dark period improved welfare of broilers

Ai Ohara[1], Tetsuro Shishido[1], Seiji Nobuoka[2] and Shusuke Sato[1]
[1]Tohoku University, 232-3 Naruko-onsen, Yomogita, Osaki, Miyagi, 989-6711, Japan, [2]Tokyo University of Agriculture, 1737 Funako, Atsugi, Kanagawa, 243-0034, Japan; oharala@gmail.com

Unhusked rice feeding is focused contribution of self-sufficiency ratio, and without influence on productivity for broilers. 51 Japanese broilers (Tatsuno) were evenly allocated into three treatments, CL (conventional feeding (C) under continuous lighting (L)), RD (unhusked rice feeding replacing 50% of corn in C (R) under 20L:4D (D)), and CD. At 5- and 7-weeks old: (1) maintenance behaviors such as feeding, drinking, sit-resting, stand-resting, moving, preening, and ground pecking of all birds were scanned at 5 min intervals for the first 30 min of each 3 h period for lighting of 20 h; (2) comfortable behaviors such as preening, dust bathing, and lateral lying of all birds were recorded by 1·0 sampling for 1 min totaling 48 minutes for each pen; (3) plumage cleanliness, hock burn, and footpad dermatitis were monitored for 10 birds in each pen. Health assessments were conducted according to Welfare Quality* Assessment protocol. Multiple logistic regression was used for maintenance behaviors analysis, and χ^2 test was used for others. (1) R and D had significant effects on the percentage of birds performing in each maintenance behavior ($P<0.01$). The percentage of birds (%, Mean±S.D) performing moving, preening, and ground pecking behaviors were more in R (2.4±1.9, 4.8±3.8, 1.4±1.7, in each behavior) than C (2.1±1.9, 4.3±3.7, 1.3±1.6), and more in D (2.2±2.0, 4.6±3.8, 1.5±1.6) than L (1.9±1.6, 3.7±3.4, 1.0±1.5). (2) Comfortable behaviors were more performed in RD than CL and CD ($P<0.05$). 3) Health assessments were better in RD than CL and CD ($P<0.05$). Unhusked rice feeding with the dark period setting may promote activity and improve welfare of the Japanese broilers.

Research on animal-assisted intervention procedures for individuals with severe developmental disorders I

Itsuko Yamakawa, Takayuki Horii, Kazue Akabane, Kanako Tomisawa and Toshihiro Kawazoe
Yamazaki Gakuen University, Department of Animal Nursing, 4-7-2, 192-0364 Minamiosawa
Hachioji-shi Tokyo, Japan; yamakawa@yamazaki.ac.jp

For individuals with severe developmental disorders to live in care facilities, they must learn to obey rules. However, this is not easy for those whose ability to understand is limited. In order to prevent accidents, their actions are restricted, unfortunately resulting in some individuals refraining from voluntary action. Partnering these individuals with a dog during the activity allows them to instigate appropriate voluntary actions towards the dog. This may eventually lead them asking appropriate voluntary actions towards human beings, indicating the importance of focusing on the approach when partnering an individual with a dog. We focused our research on whether it was preferable to approach a dog from its front or from its back. The subjects were eight residents of a care facility for adult severe developmental disorders in a suburb of Tokyo. A medium-sized dog held by a handler was used for the research. We began by approaching from the front of the dog, then from its back. Results were obtained by analyzing a video recording of these interactions. It took an average of 29.9 seconds for the individual to touch the dog when the dog was approached from the front, while it only took 1.7 seconds from its back. They kept touching the dog for an average of 2.6 seconds from the front, but 16.1 seconds from its back. In conclusion, touching the dog from the front was easier than from the back. It is indicated that they feel difficulty to accept unexpected actions of the dog, as well as the fact that the head is smaller than the back. However, in current animal-assisted activity, we often see the approach from the front of the dog. It is observed that the interaction should be begin from the front and then from the back of the dog. This research was conducted with authorization from the Yamazaki Gakuen University Ethics Committee.

The relevance to visual, auditory, and olfactory cues in pet-dogs' cognition of humans

Megumi Fukuzawa and Marina Watanabe
Nihon University, College of Bioresource Sciences, Animal Science and Resources, Kameino 1866, 252-0880, Japan; fukuzawa.megumi@nihon-u.ac.jp

There are many studies that look at how dogs respond to human communicative information; e.g. the dogs could distinguish the human scent, and discriminate the subtle changes of human verbal cues. It is also suggested that dogs might recognise such categorical transformations of human faces. However, there are no reports that there is a relative dominance in the sensory in dogs when they respond to human communicative cues. The aim of this study was to determine which human communicative factors influence the dog's response. Eleven healthy pet-dogs that didn't appear to show any aggressive behaviour toward people were recruited, 6 females and 5 males, with an age mean of 29±8.24 months. Five sensory conditions (all cues were presented, one of visual, auditory, or olfactory cue was presented, all cues were not presented) were provided three times respectively for each dog during the tests. A man was sitting on the floor in a fixed area in a box during all of the tests. If a dog's behaviour tried to confirm the man in the box was observed, the dog received a treat. All tests were video recorded, and both the behaviour and the reaching time to the man who was equipped with one of five conditions were observed. The total reaching ratio of each sensory condition in dogs was as follows; 97.0% (all cues condition), 87.9% (auditory condition), 84.4% (visual condition), 84.4% (olfactory condition), and 69.7% (no cues condition). The time that a dog could notice and then get a reward from the man who was sitting at the fixed area in the box differed between five conditions; the time of both all cues condition (6.00±0.32 s) and visual condition (6.02±0.91 s) were significantly faster than auditory (18.56±9.57 s) and no cues condition (26.55±11.72 s) (Tukey, $P<0.05$). There was a significant difference for only 'sniffing' behaviour before receiving a reward. That was observed often in the no cues condition more than all cues, visual, and olfactory conditions (Tukey, $P<0.05$). These results indicate that the information is important when a dog recognises a man, although it is related to the response time of whether a dog could confirm the aspect of the man visually.

The affiliative behavior of the dog during reunion with the owner correlated with the owner's impression in a questionnaire

Chie Mogi, Chiho Harimoto, Chie Daikuhara, Madoka Tsuya, Atsuko Oguri and Tomoko Ozawa
Yamazaki Gakuen University, Department of Animal Nursing, 4-7-2 Minami-osawa, Hachiouji,
Tokyo 192-0364, Japan; c_mogi@yamazaki.ac.jp

Dogs show affiliative behavior toward their human companions. In particular, family dogs live in a mutual attachment relationship with their owners. Dogs that failed to establish a sufficient relationship with their owners in the early developmental period, sometimes shows problematic behavior such as excessive aggression and separation anxiety. However the relevant degree of their attachment has not been investigated in behavioral tests. In this study, we used a questionnaire (with 19 questions) to investigate the attachment relationship in a sample of family dogs and conducted a behavioral test in which we observed the affiliative behaviors of dogs during reunion with the owner after a short separation. The aim of this experiment was to find out whether there was correlation between owners' impressions and affiliative behavior. Ninety five dogs were assessed by their owner, on their proximity-seeking and aggressive behaviors in their household. Fifteen dogs then served as subjects and their behavioral expressions observed in a separation test. First each dog was singly housed for 30 minutes in a cage (630×840×705 mm). During this period the dog could neither see their owner nor hear their voice. Behavioral observations were then conducted for 3 minutes to see the dog's affiliative behaviors toward its owners by placing them in an arena (2,400×2,400 mm). Conducting a factor analysis with a promax rotation of the items on the questionnaire, three factors were extracted: attachment; aggression; and noise sensitivity. The questionnaire had high levels of internal consistency as measured by Cronbach's alpha (0.737 for factor 1, and 0.772 for factor 2, only one question was extracted for factor 3). In the behavioral experiment, high correlations were found between the average score of the items of factor 1; attachment and the total number (r=0.704, $P<0.01$) as well as the total duration (r=0.682, $P<0.01$) of 'looking at the owner'. This experiment showed that measurements of dogs' behavior in a simple experiment can indicate the relevant degree of their attachment with the owner. This study was conducted with authorization from the Yamazaki Gakuen University Ethics Committee.

Hebb-Williams mazes with wild boars

Soichiro Doyama[1], Yusuke Eguchi[1], Hironori Ueda[1], Katsuji Uetake[2] and Toshio Tanaka[2]
[1]NARO Western Region Agricultural Research Center, Laboratory of wildlife management, Yoshinaga 60, Ohda city, 694-0013 Shimane, Japan, [2]School of Veterinary Medicine, Azabu University, Fuchinobe 1-17-71, Sagamihara city, 252-5201 Kanagawa, Japan; doyama@affrc.go.jp

Crop damage caused by Japanese wild boar (*Sus scrofa leucomystax*) has been increasing and it is now a serious problem in Japan. To obtain more knowledge about the wild boar is necessary to establish appropriate methods in damage control. The purpose of this study was to acquire knowledge about the spatial ability of wild boar by using the Hebb-Williams Mazes. Five captive-raised wild boars were tested. A Hebb-Williams maze for wild boars was built and it was a square 6×6 m with 1.2 m walls in an outdoor experimental field. After the training period, wild boars performed 12 test mazes with eight trials per maze. The pathway from the start to the goal of a maze and the length of time to finish a maze were recorded in each boar. Three wild boars completed all tests, and it indicated that wild boars learned and understood the spatial relations between the interior walls and the goal point with the food reward. Total Error Score of Test 12 was significantly higher than other problems (*P*<0.05) and Test 8 was the second highest. It is considered that both tests might be difficult for the large mammals to finish because similar results have been shown in the previous studies with horses and cows. Comparing maze scores of wild boars to those of other large farm animals in the previous studies, every score of wild boars showed similar or better. These results suggested that wild boars, the one of wildlife species had superior spatial ability to learn. Some maze patterns are still confusing wild boars while they are showing high learning ability of the spatial arrangements between a food and barrier fences. These could be clues for novel layouts of fencing for controlling crop damage caused by wild boars.

Influence of the distance between wire mesh and food on behavior in masked palm civets (*Paguma larvata*)

Chihiro Kase[1] and Yusuke Eguchi[2]
[1]Chiba institute of science, Shiomicho 3, Choshi, Chiba, 288-0025, Japan, [2]National Agriculture Research Center for Western Region, Yoshinaga 60, Kawai, Oda, Shimane, 694-0013, Japan; ckase@cis.ac.jp

Fencing is one of the effective measures to prevent agricultural damages caused by wildlife. Generally, it reduces animals' intentions to intrude into fields especially when the fences are away from crops. In this study, we changed the distances between wire mesh and rewards and compared how masked palm civets (*Paguma larvata*) behave by differences in it. Eight adult masked palm civets were used. The experimental room was divided into two rooms by a plywood board with an entry point at the ground level and the food as a reward installed behind the plywood board. For the tests, the entry point was covered with the 5 cm of iron wire mesh and the reward was installed at four various distances from the entry: 0 cm, 20 cm, 40 cm, and 60 cm. The duration of each trial was for 15 minutes a day each civet. The behaviors of civets were classified into three categories: Exploratory Behaviors (sniff, staring, muzzle inserting, etc.), Pushing (pushing or pressing the fence with their forepaws standing on their hind legs), and Climbing (climbing up the wire mesh). At the interval of 0 cm, as all civets easily got rewards, their actions and intentions were gone immediately. Over the distance of 20 cm, although no civets got rewards, they showed their interests during the trials. Civets spent more time showing exploratory behaviors at the distance of 20 cm than at that of 60 cm ($P<0.05$). It indicated that their motivations to take action to get rewards were reduced depends on the distances. However, for the duration of Pushing and Climbing, no significant differences were shown between the conditions and it is considered that these behaviors might not to be affected by the distances. This result suggests that distance between fences and crops will be an important factor affecting the reduction of crop damages from civets in a local farm situation.

Motor laterality in the domestic cat

Louise J. McDowell, Deborah L. Wells and Peter G. Hepper
Queen's University, Belfast, Psychology, School of Psychology, Queen's University Belfast, BT7 1NN, Northern Ireland, UK, United Kingdom; lmcdowell17@qub.ac.uk

In comparison to other species, lateralised behaviour in the felids has received little empirical attention. Studying the paw preferences of an animal is worthy of investigation, providing information on cerebral functioning, vulnerability to stress and poor welfare. This study therefore examined paw use in 50 mixed breed domestic pet cats in order to establish whether or not this species shows asymmetrical motor behaviour. Subjects were required to undertake two tasks designed to assess paw preference. The tasks differed systematically in terms of complexity. Task 1 was deemed a 'simple' task and required the animal to reach for a toy dangled above its head. Task 2 was considered a 'complex' task, requiring the cat to retrieve food from a maze by placing its paw through a small opening. Each cat performed 100 trials (i.e. paw responses) on each task. A cat handedness score was calculated in line with previous work in this area, enabling animals to be classified as left/right-pawed or ambilateral. A series of ANOVAs were subsequently carried out on the handedness scores to determine whether the direction or strength of the cats' paw use differed between the tasks, or varied according to the animals' sex (male; female) or age (young [1-6 years]; old [7+ years]). Results revealed that the direction of the cats' paw preferences did not differ significantly ($F[1,48]=1.57, P=0.22$) between the two tasks, although animals showed a significantly ($F[1,48]=409.91, P<0.001$) stronger motor bias on the more complex Task 2 (mean paw strength score = 0.58 ± 0.02) than Task 1 (mean paw strength score = 0.08 ± 0.01). Analysis revealed a significant effect of feline sex on the direction of their paw use ($F[1,48]=11.401, P=0.001$). On Task 2, significantly ($t=3.38, P=0.002$) more males (n=17, 68%) used their left paw (males mean HI=0.17 ± 0.12) while females (n=21, 84%) were more inclined to use their right paw (females mean HI=-0.35 ± -0.09). The age of the cats was not significantly ($P>0.05$) related to either the direction or strength of paw use. In sum, the findings suggest that the domestic cat shows paw preferences which centre heavily around the animals' sex. Feline paw preference also appears to be task-specific with more complex manipulatory tasks resulting in a more obvious lateral bias, a finding that may be related to functional brain specialisation.

Case report: reintegration of an adult chimpanzee with amputated left arm into social group – any impact on their behaviors?

Yoko Sakuraba[1,2], Yuji Kondo[3], Koyo Yamamoto[3], Ikuma Adachi[2] and Misato Hayashi[2]
[1]*Japan Society for the Promotion of Science, 5-3-1, Kojimachi, Chiyoda-ku, 102-0083, Tokyo, Japan,* [2]*Primate Research Institute, Kyoto University, 41-2, Kanrin, Inuyama city, 484-8506, Aichi, Japan,* [3]*Nagoya Higashiyama Zoo, 3-70, Higashiyama-motomachi, Chikusa-ku, 464-0804, Nagoya, Japan; sakuraba.yk30@gmail.com*

In December 2012, an adult female chimpanzee, named Akiko, living in Higashiyama Zoo suffered from necrosis in the muscle of her left forearm caused by an infection. She got surgery to amputate her left forearm by zoo veterinarians. After her recovery in May 2013, she was reintroduced into her group. It is important to evaluate their behaviors to provide better care not only to the handicapped chimpanzee but also to the other individuals in the same group. We observed their behaviors after her reintroduction in 2013 and compared them with their behaviors previously recorded when Akiko was intact in 2010. The observation method was the same across the two periods, a scan sampling every one minute between the hours of 14:00 h to 15:00 h which is few effect of caregivers. Sampling days were 18 days in 2010 and five days in 2013. Subjects were two adult male chimpanzees and four adult females including Akiko. The recorded behaviors were categorized as; Feed, Rest, Move, Social behaviors and Other. In addition, 'Social behaviors' were divided into affiliative / aggressive / negative behaviors to be clear what behaviors Akiko's disability impacted on. In the results, 'move' and 'social behaviors', especially affiliative behavior, decreased in Akiko (Mann-Whitney U test). However, other individuals did not show behavioral changes except one adult female. Moreover, the average of grooming time given to Akiko did not change significantly. It is suggested that the amputated arm led to difficulty of her movement and social interaction such as grooming to other chimpanzees. However, group members might not care about her handicap, since there was only a little change on their behaviors and they continued grooming Akiko. It can be one of reports to encourage others to reintroduce chimpanzees with disabilities into social group and enhance their well-being.

Comparison of ultrasonic speaker and laudspeaker for broadcasting crow vocalizations for control of crow behavior

Naoki Tsukahara[1] and Susumu Fujiwara[2]
[1]*SOKENDAI, Shonan village, Hayama, Kanagawa, 240-0193, Japan,* [2]*Mitsubishi Electric Corporation, 5-1, Ofuna, Kamakura, 247-0056, Japan; tsukahara_naoki@soken.ac.jp*

Sound broadcast from an ultrasonic speaker, which has greater directivity, can maintain volume over long distances. Broadcasting crow vocalizations using an ultrasonic speaker may be able to control the behavior of distant crows. In this study, we compared crow behavior after broadcasting crow vocalizations from an ordinary speaker and an ultrasonic speaker. The contact call (Contact) and fight call (Fight) of jungle crows (*Corvus macrorhynchos*) were broadcast from each speaker. Contact is the vocalization for contact with other crows, and is a common call. Crows use Fight when fighting with raptors, and it is a stress call. Each vocalization was broadcast using a digital audio player (iPod Nano, Apple Inc.) from an ultrasonic speaker (prototype) or a laudspeaker (RV-NB90-B, JVCKENWOOD Corporation). The calls were tested in five areas (Hayama, Fujisawa, and Yokohama in Kanagawa Prefecture, and Moka 1 and Moka 2 in Tochigi Prefecture). Crow behavior was observed near the time of call broadcast, and recorded using a drive recorder (HDR-201G, Comtec). In areas with low (35 dB(A)/15 s) and high (60 dB(A)/15 s) levels of environmental noise, crows within approximately 30 m and 10 m, respectively, reacted to the vocalizations from the laudspeaker. In contrast, when the ultrasonic speaker was used in areas with low and high levels of environmental noise, crows within approximately 100 m and 50 m, respectively, reacted to the vocalizations. Thus, use of an ultrasonic speaker to broadcast crow vocalizations could be used to control crow behavior over greater distances than a laudspeaker.

The effects of wolf's feces and dried starfish on avoidance behaviour in Sika deer (*Cervus nippon yesoensis*)

Naoshige Abe and Hitoshi Ogawa
Tamagawa University, agriculture, 6-1-1 Tamagawa-gakuen Machida-shi Tokyo, 194-8610, Japan; naoabe@agr.tamagawa.ac.jp

Wild unglates might inflict serious feeding damages to forest and agricultural products. The purpose of this study was to evaluate the avoidance effects of wolf's (*Canis lupus*) feces (Wf) and dried starfish (*Asterias amurensis*) on Sika deer (*Cervus nippon yesoensis*). In experiment 1, the number of Sika deer moving to the grassland from the forest was observed before and after exposure to wolf's feces (Wf) on the migration route. The number of Sika deer was recorded for 20 days in Sept.-Oct. 2012 using video (13 m height). On day 18th Sept.,ten feces samples (50 g) were placed in bottles (diameter, 8 cm) at intervals of five meter on the grassland. The differences in number of deer were examined with one-way ANOVA and Tukey's test. Sika deer appearance was reduced after exposure to Wf (pre showing 18.9, postshowing 7.8; $P=0.055$). Furthermore, the number of Sika deer increased significantly (to 21.4; $P<0.01$) 3 weeks later after heavy rain (9 mm/h). In experiment 2, dried starfish (SF) was placed on the migration route (40 kg SF were spread in a 30 cm band for 40 m). The observations were carried out for 29 days in May-June 2014 before and after SF was spread on day 25th May. Sika deer appearance was significantly reduced after expose to SF (pre showing 34.5, post showing 13.7; $P<0.01$). However the number of Sika deer increased to 29.9 4 week s later ($P<0.01$) following the first heavy rain(15 mm/ hour). In conclusion, these results suggest that Wf and SF induce avoidance behaviour in Sika deer for about one month until heavy rain.

Genome-wide association study for different indicators of cattle temperament using the single step procedure (ssGBLUP)

Tiago S. Valente, Fernando Baldi, Aline Sant'anna, Lucia Albuquerque and Mateus Paranhos Da Costa

Faculdade de Ciências Agrárias e Veterinárias, UNESP, Jaboticabal-SP., Departamento de Zootecnia, 14884-900, Brazil; tiagosv_bio@hotmail.com

The aim with this study was to identify overlap of genomic regions associated with different traits of Nellore cattle temperament, which was assessed during the yearling weight procedure. The behaviour of each individual was scored when inside the crush, measuring: (1) movement (MOV), scoring (from 1 to 5) the cattle displacement; (2) tension (TENS), scoring (from 1 to 4) the body tension, and head, ear and tail movements; and (3) crush score (CS), which assigns scores from 1 to 5 to assess the overall cattle reactivity. Data were from 16,119 animals with phenotypes for each temperament trait and pedigree file with 162,644 animals. A total of 1,405 animals were genotyped with BovineHD BeadChip. The quality control (QC) was performed to exclude SNP markers with unknown genomic position, located on sex chromosomes, monomorphic, MAF<1%, call rate<90%, and animal call rate (with less than 90% of SNPs called). After QC, 455,374 SNPs and 1,384 genotyped animals remained. The association analyzes were performed using ssGBLUP, a modification of BLUP with numerator relationship matrix A^{-1} matrix replaced by H^{-1}, that uses the GEBV to estimate the SNP effects. The variance components and genetic parameters were estimated by Bayesian inference via Gibbs sampling, assuming a threshold models for MOV, TENS and CS, which included direct additive genetic and residual effects as random effects and contemporary groups (farm and year of birth, sex and management groups at birth, weaning and yearling) as fixed effect. The SNP effects were calculated to segments of 10 sequential SNPs and results interpreted as the percentage (%) of total genetic variance explained by each window. The segments that explained 1% or more of the total genetic variation were considered as a genomic region associated to each temperament traits. A total of 18, 12 and 10 windows associated with MOV, TENS and CS, respectively, were identified. The overlap regions between MOV and TENS were on BTA4 (4:112163473-112184024, 1.27%) and BTA17 (17:69709764-69739635, 2.41%), and between TENS and CS was on BTA24 (24:51252080-51265610, 2.77%). No candidate region was associated with MOV and CS simultaneously. This result confirms the interaction between different genes to modulate the expression of cattle temperament. These candidate regions associated to distinct traits can be used for future studies in order to identify causal genes affecting overall aspects of beef cattle temperament.

A study on validity and reliability of a temperament test for Japanese pet dogs

Kana Mitsui

Teikyo University of Science, Graduate School of Science & Engineering, 2-2-1 Senju-sakuragi, Adachi-ku, Tokyo,120-0045, Japan; kana_mitsui0718@yahoo.co.jp

The temperament test for dogs can play a significant role for adopters to help to know their new dogs better before they start living together. Although such tests are now being implemented at many animal shelters, there are few studies on validity, and reliability of those tests. The purpose of this study was to conduct a temperament test with Japanese pet dogs to see their characteristics and to examine the reliability of the test when repeatedly conducted with two evaluators. Eighteen dogs (>1 year old) were assessed with the ASPCA/NEVBA behavioural test by two evaluators. The dogs were videotaped during the test and the same test was repeated two weeks later. There were 9 items in the test;Room behaviour, Leash manners and Commands, Handling, Response to toys,Run and freeze,Response to foods,Approach by toddler doll,Behaviour with strange persons,and Behaviour with another dog. Two evaluators put scores from 0 to 3, for each of four types of nature (friendliness, fear, arousal, aggression) for each test item by observing behavioural signs of dogs using the standardised scale for interpretation. Along with this subjective assessment, behavioural analysis was carried out to record the frequency and duration of 16 items of vocalisation, posture, and behaviour using the video. Factor analyses were conducted for an evaluator's scores for each time of the tests, and the results of behavioural analyses were compared to check the consistency by Wilcoxon signed-rank test. Six factors were extracted for each factor analyses and accumulated contribution rates were over 0.13. Three factors were common in the results for one evaluator and only one factor was common for the other between the first and second test. Two evaluators agreed three factors for the first test and four factors for the second test. Behavioural analysis did not show significant differences in 10 items ($P<0.05$) between two tests. Those behavioural responses observed consistently were barking, growling, play-bow, tail tucked, lying down, jump on object, show belly, lower body, withdraw, and hiding. In conclusion, although there was moderate level of consistency in the behavioural test when repeated, the intra- and inter-evaluator reliability was not very high. These results suggest that the evaluators should be aware of the behavioural responses consistent over time and they may need to observe such behaviours for more reliable assessment.

Does the evolution of parental care in decomposer affect the decomposition efficiency of dead animals?

Mamoru Takata

Tokyo University of Agriculture and Technology, Graduate School of Agriculture, 183-8509, 3-5-8 Saiwai, Fuchu, Tokyo, Japan; mamorururu2000@yahoo.co.jp

Decomposition of dead animals is an important process for sustainable society in terms of the material circulation and the prevention of disease. Decomposition speed is dependent on the species of decomposer, however, it is currently unknown what kind of biological deference in decomposer affect the speed. In this study, I focused on the biology of two burying beetle species, Nicrophorus quadripunctatus and Ptomascopus morio. These species are major decomposer of dead animals in Japan. Both N. quadripunctatus and P. morio provide care for their offspring (ex. prepare food resources for their larvae and guard them from competitors). The major biological difference in these species is presence of food provisioning. N. quadripunctatus provisions predigested carcass to their larvae, whereas P. morio does not. The parental food provisioning may speed up decomposition by increasing growth rate of their larvae. To investigate the influence of parental food provisioning on the decomposition efficiency, I investigated whether parental food provisioning increases growth rate of their larvae. Firstly, I compared the growth rate of larvae between these two species. Secondly, I compared the growth rate of larvae in N. quadripunctatus between when larvae were provisioned by parent and when larvae grew without provisioning. To investigate the difference in growth rate, we used a generalized linear model (GLM). The larval body weight at 120 h old was treated as a response variable assuming a Gaussian distribution, and the experimental groups was treated as explanatory variables. Data were analysed separately for the first and second experiments. All GLMs were conducted using R 3.1.1 GUI 1.65. P-values were calculated using the likelihood ratio test. The body weight of larvae with parental food provisioning were significantly heavier than the larvae without provisioning (between species, estimate = -192.509, d.f.=1, 84, $P<0.001$; within species, estimate = -168.058, d.f.=1, 76, $P<0.001$). These results suggest that parental food provisioning increases decomposition efficiency by increasing the larval growth rate.

Free-roaming cat populations in Onomichi City, Hiroshima, Japan: comparison between uptown and downtown areas

Aira Seo and Hajime Tanida

Hiroshima University, Graduate School of Biosphere Science, 1-4-4 Kagamiyama Higashi-Hiroshima, 7398528, Hiroshima, Japan; d134763@hiroshima-u.ac.jp

The free-roaming cats that inhabit both the uptown and downtown areas of the old district of Onomichi city attract many tourists. The uptown area is characterized by residences, temples, souvenir shops, restaurants and cat museums which are all located along the steep slope of the Senkoji mountain park leading to the Senkoji temple on the mountaintop. The tourist map for the park shows the spots where tourists can meet free-roaming cats. The downtown area facing Onomichi Channel is composed of general stores, souvenir shops (selling souvenirs featuring cats), restaurants and bars located along the shopping arcade. However, local residents regard the growth of the cat population as a problem mainly from a hygienic point of view, and yet a study on the free-roaming cat population in this area has not been conducted. This study compared free-roaming cat populations found in these two areas of the old district. Cat population was determined using the route census technique in both areas. The total distance of the route was 3 km. The observation was conducted from 10:00 to 16:00 twice a week in both areas. The starting point was randomly chosen at each observation. The pilot observation was carried out for a total of 75 h from May 2011 to January 2012. The main survey was then conducted for a total of 864 h from February 2012 to January 2015. Each cat was recognized with coat color pattern, size, hair length and the shape of the tail. The cats were all photographed and the observed time and points were recorded. The numbers of cats identified in the uptown and downtown areas were 207 and 141, respectively, during the entire survey period. The number of cats observed in the uptown area somewhat decreased in 3 years, but the number in the downtown area slightly increased in 2014. The cat's turnover rate for each population was 43.9% (uptown) and 44.0% (downtown). The annual distribution of the numbers of cats identified was not significantly different between the areas ($P=0.16$, Chi-square test). There were 2 peaks (June and October to January) in the number of cats identified in both areas. The observation interval for each cat (the average period that a cat was observed again) was significantly shorter in the uptown than the downtown area ($P<0.05$, ANOVA). In conclusion, the demographic character of the free-roaming cat populations in both areas is different. The cat population in the uptown area has a tendency to depend on feeding by tourists, but residents are the main feeders for cats in the downtown area. The study suggests that any proposed control method of free-roaming cat populations should be designed separately between the areas.

Short-term memory testing using colour vision in laying hens

Yumi Nozaki[1], Yuko Suenaga[1], Shuho Hori[2], Chinobu Okamoto[1], Ken-ichi Yayou[3] and Shuichi Ito[1]

[1]Tokai University, School of agriculture, Kawayo, Minamiaso-mura, Aso-gun, 869-1404 Kumamoto-ken, Japan, [2]Tokai University, Graduate School of agriculture, Kawayo, Minamiaso-mura, Aso-gun, 869-1404 Kumamoto-ken, Japan, [3]National Institute of Agrobiological Sciences, Tsukuba-shi, 305-8602 Ibaraki-ken, Japan; xiahy0seop62nz@i.softbank.jp

This study consisted of four experiments using a two-choice delayed-response test to investigate whether laying hens could recall the location of a coloured card. We used two commercial strains of laying hens, four Julia and five Boris Brown hens. A red card and an achromatic colour card were used as a presentation stimulus. The positive stimulus was the red card, and the negative stimulus was the grey card. Furthermore, two white cards were prepared to hide coloured cards. Hens were investigated using operant conditioning tests using a Y maze. First, hens were trained to learn the relationship between a positive target (coloured card) and a reward (feed) and to discriminate between red and grey cards. In experiment 1, red and grey cards were shown to hens. Then, they were made to select an arm of a Y maze after hiding these cards with a white card. Experiment 2 was set up in such manner so that hens could react earlier than they reacted in experiment 1. In experiment 3, coloured cards were not hidden by a white card but the same red card was used as a positive stimulus. Finally, in experiment 4, we used new, untrained hens and exposed them to the same procedure as that in experiment 1. All experiments consisted of 20 trials per session. The criterion of a successful discrimination test was two consecutive sessions with more than 16 correct choices ($P<0.05$, Chi-square test). None of the Julia hens were able to learn the relationship between the positive stimulus and the reward. Two of the Boris Brown hens were able to learn this relationship. However, neither of these two hens achieved the criterion in experiment 1 but both of them achieved the criterion in experiment 2. The hens seemed to have decided already advancing direction from the previous the card was hidden in experiment 2. In this case, it was thought that hens did not use their memory. Furthermore, the criterion for successful discrimination was not achieved in either experiments 3 or 4. It remains challenging to estimate the short-term memory capabilities of hens from our results. It might be difficult for laying hens to study the delayed-response test using a colour stimulus. Alternatively, laying hens might not have short-term memory capabilities for recalling the location of a coloured card.

Numerical discrimination of visual stimuli by pigs: two versus three

Ryousei Ueno[1] and Tohru Taniuchi[2]
[1]Ishikawa Prefectural University, Farm, 1-308 Suematsu Nonoichishi Ishikawaken, 921-8836, Japan, [2]Kanazawa University, Graduate School of Socio-Environmental Studies, Kakumamachi Kanazawashi Ishikawaken, 920-1192, Japan; ueno@ishikawa-pu.ac.jp

Previously we studied simultaneous numerical discrimination by pigs of two and three dots and found that pigs utilized the location of a dot in a specific position in a LCD monitor, possibly because the probability of a dot in that specific location was inevitably higher with three dots sets than that in sets of two dots. In the present study in order to reduce effectiveness of such local cues, we examined similar discrimination learning of two versus three dots using animation movies. The subject was one castrated pig, approximately eight years old. Discriminative stimuli consisted of two or three circular dots. The dots were blue and 6 cm in diameter. Adjacent two LCD monitors presented either of the stimulus sets, which consisted of two or three dots. These stimulus sets rotated clockwise in the monitors. Nose poke responses to the LCD with the correct stimulus, that is, three dots, were reinforced by a 10 g apple piece. A 48 trial session was conducted daily. The subject attained above 80% correct performance after 22 sessions. Static stimulus sets were inserted occasionally to compare with our previous finding of usage of the local cue. The subject also showed good performance to the static stimuli as well as to the rotating stimuli and close analysis showed the local cue was not used. The subject also maintained good performance across tests where the dot size was changed to assess the gross area of the dots in stimulus sets as a possible discriminative cue. These results strongly suggest that pigs can learn abstract numerical properties of visual stimuli.

Memory-based performance in radial maze learning in tortoises (*Agrionemys horsfieldii*)

Tohru Taniuchi
Kanazawa University, Kakuma-machi, Kanazawa, Ishikawa, 920-1192, Japan;
tohruta@staff.kanazawa-u.ac.jp

Central Asian tortoises (*Agrionemys horsfieldii*) were trained in an eight-arm radial maze surrounded by rich extra-maze cues. Food rewards were 300 mg green vegetables and presented in food cups of 1 cm depth. Firstly, tortoises were trained in a free-choice task in which they were allowed to enter into the arms based on their own choice. Training was conducted one trial per day. After 80-140 trials of the free-choice task, tortoises showed approximately 7.0 correct responses in the first eight choices, reliably above chance level. However, they developed a strong tendency to select arms in a rotating manner and selected adjacent arms more than 50% against chance level (12.5%). Such kinds of stereotypic response patterns enable animals to get food rewards without them retaining in working memory which arms were visited. Therefore, the task was shifted to forced- and free-choice task where the order of the first four choices was decided by the experimenter so that tortoises were required to select four unvisited arms among the eight arms. Thus, stereotypic response patterns could not contribute to improvement of performance in the forced- and free-choice task. Tortoise showed 665-70% correct responses in this task reliably above chance (50% for without replacement). An additional test suggested tortoises did not utilize possible visual or olfactory cues of food rewards. The results expand a cross-species generality of previous findings that showed radial maze learning by red-footed tortoises, suggesting that tortoises have a working memory process which they utilize to guide complex spatial behavior.

Selectivity for rangeland characteristics by grazing yaks in Western Nepal, Himalaya

Hiroki Anzai[1], Manoj Kumar Shah[2], Mitsumi Nakajima[1], Takashi Sakai[1], Kazato Oishi[1], Hiroyuki Hirooka[1] and Hajime Kumagai[1]
[1]Kyoto University, Graduate School of Agriculture, Kitashirakawa-Oiwakecho, Sakyo-ward, Kyoto 606-8502, Japan, [2]Nepal Agricultural Research Council, Livestock Product Production and Management Regional Agricultural Research Station, Lumle, Kaski, Nepal; anzai@kais.kyoto-u.ac.jp

Yaks (*Poephagus grunniens*), major livestock in the Himalayas, are raised by year-round grazing with seasonal migration. Although yak grazing plays a key role in maintaining the rangeland in the region, its effect has not yet been quantified. For appropriate rangeland management, it is important to understand the effect of land cover types and topography on the land use by grazing yaks. The objective of this study was to analyze the selectivity for the rangeland characteristics by grazing yaks using information from GPS and GIS. The positional data of three yaks reared in the south of Mustang District, Nepal were collected every 10 minutes from October 2013 to March 2014 using GPS collars. The GPS data were divided into eight periods, and the data obtained from 10 am to 3 pm were extracted as the free-ranging observed data. The 95% kernel home range for each period was estimated by the fixed kernel method as the grazing area. Ivlev's electivity index was used to determine the preferences by comparing rangeland characteristics of the observed data points and those of the estimated grazing area. The preferences for land cover types, slope gradients and slope direction were analyzed in this study. The results showed that the yaks significantly preferred grassland to forest and bare land. Regarding the difference in slope gradients, the yaks significantly avoided the rangeland with the gradient more than 50° compared with the lower gradient. As for slope direction, north- and southwest-facing slopes were significantly avoided by the yaks. The information on the selectivity for the rangeland characteristics obtained in the present study will be useful to estimate grazing distribution and pressures of yak herds.

The changes of grazing time, moving distance and the distance between individuals for Thoroughbred foals until weaning

Tomoki Tanabe[1], Masahito Kawai[1] and Fumio Sato[2]

[1]Obihiro University of Agriculture and Veterinary Medicine, Inada-cho, Obihiro, 080-8555 Hokkaido, Japan, [2]Japan Racing Association, Hidaka Training and Reserch Center, Urakawa-cho, Hidaka, 057-0171 Hokkaido, Japan; s26288@st.obihiro.ac.jp

To investigate the changes, according to age, in grazing and resting time, moving distance and the distance between individuals, 6 Thoroughbred foals and their broodmares on grazing pasture in a private breeding farm were used from 1 to 6 months of age for foal. Once a month, a portable GPS was fitted on the head-collars of all the horses to record moving distance and the distance between individuals. These sessions were done for a period of 600 minutes from morning to evening from 1 to 2 months of age and 1000 minutes from afternoon to the next morning for 3 to 6 months of age for the foals, respectively. In addition, the behavior was visually checked at 5 min intervals for a 4-hour period starting 30 minutes after being grazed on the pasture. The total time spent for grazing and resting were calculated based on the observation records and the GPS data. The data were analyzed statistically by ANOVA, Student t-test and Steel-Dwass. Grazing time of foals increased with an increase age and it was not different from that of their broodmares at 4 to 6 months of age for the foals, and the similar change of their resting time was observed. Average moving distance of the foals at 1 and 2 months of age was shorter than from 3 to 6 months of age (9.6 vs 11.5 km/d, $P<0.05$). Total running distances at a trot (2 to 5 m/sec.) and a canter (> 5 m/sec.) for 1-month old foals were 1,272 and 1,168 m, respectively, which were the longest of all the ages of foals. In 1-month old foals, the distance between foals and their broodmares was shorter than between foals and their nearest neighbor foals ($P<0.05$). However, at 6 months of age, the distance between foals and their nearest neighbor foals was shorter than between foals and their broodmares ($P<0.05$). In conclusion, moving and running distance results suggest that 1-month of age is the most active period for a foal. The distance between individuals data shows that 6 months of age is the time when foals begin to become more independent from their broodmares, but the grazing time of foals is similar to their broodmares at 4 months of age, which suggest that foals may be independent at 4 months in regard to nutritional intake.

Influence of season, feeding time and bedding on recumbent rest in horses

Aya Iwamoto and Hajime Tanida

Hiroshima University, Graduate School of Biosphere Science, 1-4-4 Kagamiyama Higashi-Hiroshima, 7398528, Hiroshima, Japan; aya-hippo18@hiroshima-u.ac.jp

Some studies have suggested that the sleep pattern of horses depends on age, diet, bedding and familiarity with its environment. Horses are able to sleep while standing, but they are unable to complete a sleeping cycle without lying down to enter paradoxical sleep (PS). It is thus assumed that recumbent time is an indicator of PS state, and PS is expected to be crucial for the performance of horses. However, the factors that influence recumbent rest have not yet been fully studied. The purpose of the study was to examine the influence of season, feeding time and bedding on recumbent rest in horses. The study was conducted from January to December of 2014. Five riding horses of Thoroughbred (1 mare and 4 geldings) belonging to the Hiroshima University equestrian club were used. The age of the horses were 6 to 15 years. The horses were housed in individual boxes (3.5×3.5 m) with opening windows so that they were exposed to natural climate changes such as temperature and humidity all year round. The bedding of the 3 horses was changed from wood shavings to straw in December. The horses underwent fitness training 6 days a week. They were brushed and fed a mixture of oats and a supplement, and hay. They were subjected to the same feeding (4 times a day) and management routine. The time of evening feeding was set at 21:00, but they were fed 1 hour earlier once in a while for convenience. The behavior of the horses was recorded on a time-lapse digital camera every one minute for 97 d during the study period. Seasonal effect was analyzed with data obtained from a minimum of 5 consecutive nights per month. The average percentage of recumbent rest among the horses at night (for 6 h from 22:00 to 04:00) was 20.0%. The hourly pattern of recumbent rest at night was significantly different among the horses ($P<0.05$,). There were significant differences in recumbent rest of the horses ($P<0.01$) during a 12-month period, showing a tendency of short recumbent time in summer (July to September). The time of evening feeding was either 21:00 (43 d) or 20:00 (49 d) for 92 d observation period. When the hourly pattern of recumbent rest within 6 hours after each evening feeding was compared between the 2 feeding regimens, the hourly pattern was significantly different ($P<0.05$) in three out of 5 horses. However, the duration of recumbent rest per night was not significantly different among the horses. Bedding changes from shavings to straws for 3 horses resulted in 35.6% reduction of recumbent rest in the following 5 d ($P<0.01$). In conclusion, it is possible that recumbent rest gets shorter in summer, and that the unexpected changes of the time of evening feeding and changes in bedding affect the hourly pattern of recumbent rest of the horses.

The effects of space allowance on the behavior and physiology of cattle temporarily managed on rubber mats

Karin Schütz[1], Frances Huddart[1], Mhairi Sutherland[1], Mairi Stewart[2] and Neil Cox[3]
[1]AgResearch Ltd, Ruakura Research Centre, Private Bag 3123, Hamilton 3240, New Zealand, [2]InterAg, Waikato Innovation Park, P.O. Box 9466, Hamilton 3240, New Zealand, [3]AgResearch Ltd, Invermay Research Centre, Private Bag 50034, Mosgiel 9053, New Zealand; karin.schutz@agresearch.co.nz

Dairy cows in some pasture based systems are taken off pasture for periods of time in wet weather to protect pastures. The use of rubber matting for such stand-off practices is common to improve comfort. Our aim was to study the effects of different space allowances on cow behavior and physiology when managed temporarily on rubber mats. Thirty pregnant, non-lactating Holstein-Friesian cows were divided into 6 groups of 5 and exposed to 6 space allowance treatments (3.0, 4.5, 6.0, 7.5, 9.0 and 10.5 m²/cow) on a 24 mm rubber surface during a stand-off period which consisted of 18 h in treatment pens (with no feed provided) followed by 6 h on pasture (to allow for daily feed intake) for 3 consecutive days (6 days of recovery on pasture between treatments). All groups received all treatments. Cows spent more time lying with increasing space allowance during the stand-off period (ANOVA;$P<0.001$) and less time lying on pasture during their 6 h feed break (ANOVA; $P=0.021$). Average lying times (pasture and rubber mats combined) for the different space allowances were for 3.0 m²=7.5 h, 4.5 m²=10.2 h, 6.0 m²=11.9 h, 7.5 m²=12.4 h, and for 10.5 m²=13.8 h/24 h. When cows were on recovery on pasture in between treatments they spent 11.2 h/24 h lying, on average. Aggressive interactions (ANOVA; $P=0.006$) and non-aggressive lying disturbances (ANOVA; $P=0.006$) were more frequent at lower space allowances (aggressive interactions decreased by 35% from 3.0 to 4.5 m²/cow, with a slower decline thereafter). Cows were dirtier after the stand-off period (ANOVA;$P<0.001$), particularly at lower space allowances. Cows had higher gait scores (by 0.1 unit on a 5-point scale, higher scores indicate more lameness) after the stand-off period but this change was unaffected by space allowance (ANOVA; $P=0.349$). Stride length, plasma cortisol and body weight were unaffected by the stand-off period and space allowance (ANOVA; $P\geq0.103$). The results suggest that providing more space for dairy cattle managed temporarily on rubber matting is beneficial in terms of animal welfare.

Impact of cold exposure and shearing stress during pregnancy on lamb latency to bleat

Lea Labeur[1,2], Sabine Schmoelzl[1,2], Geoff Hinch[1] and Alison Small[1,2]
[1]*University of New England, Animal Science, Animal Science, School of Environmental and Rural Science, University of New England, 2350 Armidale NSW, Australia,* [2]*CSIRO, Agriculture, Chiswick New, 2350 Armidale NSW, Australia; lea.labeur@csiro.au*

Lamb mortality is a welfare and production issue. Shearing ewes during pregnancy can increase lamb birth weight (BW) but the role of stress(shearing or subsequent cold exposure) is not fully understood. BW and bleat latency (LB) are genetically correlated to lamb survival. We examined the effect of cold stress (CS) in pregnant ewes on lamb LB when separated from the mother 4 h after birth, before and after a one hour cold challenge(CG) at 1 °C. The first experiment examined the effects of the timing of shearing of ewes (mid-pregnancy vs late-pregnancy). Sixty Merino ewes were divided into four groups balanced for pregnancy status. One group was shorn, at 90 d and one at 130 d of pregnancy. Both groups were subsequently wetted, using sprinklers, 3 times over the 10 d following shearing. One control group was sham handled simultaneously alongside each shorn group but no other stressors applied. Lambs born to mid-pregnancy control ewes had shorter LB than lambs born to mid-pregnancy shorn ewes before CG ($P=0.038$; 14.2 vs 36.7 s). The second experiment examined the effects of repeated cold exposure (CE) during the last month of pregnancy. Sixty previously shorn Merinos ewes were divided into two groups balanced for pregnancy status. During the last month of pregnancy one group was exposed to 0 °C in a controlled temperature room for 3 h on 3 occasions and the other received no cold stress. Results showed that CE had an inconsistent effect on lamb LB. Lambs born to CE ewes were slower to bleat than control lambs at 4 h after birth($P<0.01$; 7.9 vs 2.3 s) but there was no difference in LB after a lamb CG. CE and stress during pregnancy had an inconsistent effect on lamb LB. Ongoing work investigates impact on other lamb vigour traits. Non-parametric Mann-Whitney statistical tests were used. The protocols were approved by the CSIRO Armidale ethics committee (ARA14/17 and 14/30).

The effects of alternative weaning method on behaviour in swamp buffalo calves

Pipat Somparn[1], Nasroon Chalermsilp[2] and Suriya Sawanon[3]
[1]*Thammasat University, Agricultural Technology, Faculty of Science & Technology, Thammasat University, Rangsit Campus, Pathum thani, 12120, Thailand,* [2]*Thammasat University, Agricultural Technology, Faculty of Science & Technology, Thammasat University, Rangsit Campus, Pathum thani, 12120, Thailand,* [3]*Kasetsart University, Animal Science, Faculty of Agriculture at KamphaengSaen, Kasetsart University, KamphaengSaen Campus, Nakhon phatom, 73140, Thailand; somparn@tu.ac.th*

Artificial weaning of swamp buffalo calves is usually done abruptly and early compared to the natural weaning of the species. Enforced mother-calf separation is stressful for the animals. The aim of this study was to compare the behavioural responses and weight change of swamp buffalo calves weaned using two weaning methods. It was undertaken at Buriram Livestock Breeding Station, Buriram Province from 30 December 2013 to 20 November 2014. Twenty-seven swamp buffalo calves (12 females and 15 males) were blocked by season and randomly assigned within block to one of the three treatments being: (1) CON: cows and calves remained together (2) TRAD: weans abruptly on day 0; (3) ALT: cows and calves remained together but suckling was prevented on day -14 by inserting a nose flap antisuckling device, followed by remote separation and removal of the device on day 0. After separation, cows and calves from treatment 2 and 3 were completely isolated from each other, prohibiting visual contact or vocal communication. Behaviours were recorded individually by instantaneous and behaviour sampling methods on day 0 and day 7. Calf weights were recorded on day 0 and day 30 after the separation. On day 0, the TRAD calves produced 53 calls/h, which was approximately two times more than ALT calves (23 calls/h, $P<0.05$). TRAD and ALT calves walked ($P<0.05$) more than CON calves. TRAD calves spent less time eating than CON ($P<0.05$) but did not differ from ALT calves ($P>0.05$). On day 7, TRAD calves walked ($P<0.05$) more than ALT calves but did not differ from CON calves. TRAD calves spent less time eating than CON and ALT calves ($P<0.05$). Over the entire study period, ADG did not differ ($P>0.10$) for both weaning treatments but had lower ($P=0.02$) ADG than CON calves. In conclusion alternative weaning method did not appear to provide any clear benefits in reducing weaning distress particularly weight losses following separation. Further research is required if we are to understand longer-term studies of compensate for these early losses in weight gain.

Mother-offspring vocal communication in sheep and goats as an animal welfare index

Hirofumi Naekawa
Tokyo University of Agriculture, Teacher Education and Scientific Information, Atsugi-shi, 243-0034, Funako 1737, Japan; h3naekaw@nodai.ac.jp

We investigated vocal communication in sheep and goats experiencing distress related to mother-offspring separation as a useful objective index concerning animal welfare. Data were collected at a nomadic summer pasture on the Mongolian Plateau, where each offspring was isolated from their respective mothers (sheep: 617, goats: 251), and classified into four behavioral type-based categories: moving, moving/feeding, feeding, and resting. Phonation duration, basic frequency, formant, and sound pressure were analyzed using sound analysis software (ANIMO, Japan). Discriminant analysis was performed on lamb vocalization data according to behavioral type, within-herd position, and vocalization target. A significance test was then performed for linear discrimination coefficients. A significance test for linear discrimination coefficients was conducted based on goat vocalization data, following which a significance test was performed for each category based on canonical discriminant analysis. Results for lambs showed a difference in the phonation duration between one group targeting the mother and the other targeting the herd ($P=0.008$, $P<0.05$). The phonation duration and second formant for mother-targeted vocalization differed depending on the animal's within-herd position ($P<0.05$) and behavioral types ($P<0.05$). For goats, there was significant variation in the mother and kid across different categories ($P<0.01$), suggesting a possible distinction between variance-covariance matrices. The phonation duration ($P<0.05$) and sound pressure ($P<0.001$) were both significantly different in mother and kid, which suggested their contribution to the discrimination. In conclusion, acoustic parameters of offspring's vocalization toward their mothers may vary depending on phonation duration, within-herd position, and behavioral type in lambs; however, it may vary depending on phonation duration and sound pitch in goats.

Mother and partner promote exploratory behavior in *Octodon degus*

Tomoko Uekita
Kyoto Tachibana University, Psychology, 34 Yamada-cho Oyake Yamashina-ku Kyoto, 607-8175, Japan; uekita-t@tachibana-u.ac.jp

Little is known about how social interactions affect emotional states and behavior in *Octodon degus* (degus). To address this issue, we first investigated how the presence of a mother affects object exploration in 3- to 7-week-old degus. Young degus with or without a mother (no-mother and with-mother conditions) were allowed to explore a novel object presented in their home cage for 5 min. The frequency of contact with the object and the time spent reaching toward the object were analyzed using two-way ANOVA with age and condition as within-subject factors. Contact with the object was more frequent in the with-mother condition than in the no-mother condition ($P<0.05$). Time spent reaching for the object was less in the with-mother condition than in the no-mother condition at 6 weeks of age ($P<0.001$), the weaning period. These results suggest that the presence of a mother might reduce anxiety and promote novelty seeking in young degus. Next we observed foraging and vigilance behavior of 2-year-old degus. Degus were allowed to explore an open-arena scattered with sunflower seeds either alone (unpaired condition) or with a familiar same sex conspecific (paired condition), then alarm calls were played back to them. We measured durations of foraging and freezing after playing the alarm call. A Wilcoxon signed-rank test for within-subjects comparisons revealed that freezing duration in the paired condition was shorter than in the unpaired condition for male degus ($P<0.05$). A Mann-Whitney U test for between-subjects comparisons revealed that in the paired condition, male degus exhibited longer foraging behavior than female degus ($P<0.05$). These results indicate that the presence of a partner attenuates anxiety induced by alarm calls in male degus, and paired male degus were able to forage longer than paired females.

Production, behaviour, and welfare of organic laying hens kept at different indoor stocking densities in multi-tier aviaries

Sanna Steenfeldt[1] and Birte L Nielsen[2]
[1]Aarhus University, Department of Animal Science, Research Center Foulum, 8830 Tjele, Denmark, [2]INRA UR1197 NeuroBiologie de l'Olfaction, PHASE, Bât 320, 78352 Jouy-en-Josas, France; sanna.steenfeldt@anis.au.dk

Multi-tier aviary systems are becoming more common in organic egg production. The area on the tiers can be included in the usable area available to the animals when calculating the maximum indoor stocking densities. The present study investigated the differences in production, behaviour, and welfare of organic laying hens (from 18 to 65 weeks of age) in a multi-tier system with permanent access to a veranda and kept at stocking densities (D) of 6, 9, and 12 hens/m^2 available floor area (not including tiers) by increasing group size (396, 594, and 792 hens per group with 3 replicates per treatment). In addition, in a fourth treatment access to the top tier was blocked reducing vertical, food, and perch access at the lowest stocking density (D6x). In all other aspects than stocking density, the experiment followed the EU regulations on the keeping of organic laying hens. Three types of effects emerged from this comparison (GLM with repeated measures) of indoor stocking densities: Firstly, certain measurements (live weight, mortality, and foot health) were not significantly affected by the stocking densities used in the present study. No systematic effects of density were found on laying behaviour (eggs laid outside nests, nest preferences, egg weight). Secondly, some variables (laying percentage, use of veranda and outdoor area) were affected directly by the reduced floor (and nest) space per hen; e.g. laying percentage was lowest in D12 (90.6±0.7%) compared to the rest (94.3±0.7; $P=0.001$). Finally, some variables (plumage condition, presence of breast redness and blisters, pecked neck- and tail feathers, as well as perch use) were affected by the simultaneous reduction in access to other resources than floor space, mainly perches and troughs; e.g. neck feather damage and breast redness were seen more in treatments D6x and D12 than in D6 and D9 ($P<0.001$ and $P=0.036$, respectively). The welfare of the hens was mostly affected by these associated constraints, despite all of them being within the allowed minimum requirements for organic production in the EU. Although these welfare consequences were moderate to minor, it is important to consider concurrent constraints on access to other resources when higher stocking densities are used in organic production.

Effect of pre- and post-partum sow activity on maternal behaviour and piglet weight gain 24 h after birth

Gudrun Illmann, Helena Chaloupková and Kristina Neuhauserová

Institute of Animal Science, Department of Ethology, Pratelstvi 815, 10400 Prague, Czech Republic; gudrun.illmann@vuzv.cz

Within the 24 h prior to parturition sows are active and motivated to perform nest-building behaviour. The aim of this study was to investigate: (1) whether pre-partum activity (e.g. nesting and postural changes) could predict maternal behaviour 24 h post-partum (pp) and weight gain 24 h pp, and (2) whether post-partum activity (post-partum nesting and postural changes) affect parts of the maternal behaviour 24 pp (e.g. first suckling, udder access, suckling duration) and piglet weight gain 24 h pp. Fifteen sows were housed in modified straw-bedded 'walk-around' farrowing crates. Pre-partum nesting events and postural changes were recorded 24 h before parturition. During parturition the number of nesting behaviour and latency of the first sucking of the whole litter were recorded. Number of postural changes and duration of udder access were recorded 24 h after birth of the first piglet (BFP) during three time periods (during parturition, from the end of parturition to 12 h after BFP, and 12-24 h after BFP). The duration of suckling and sow responsiveness on the playback of piglets' scream were recorded during two time periods (from the end of parturition to 12 h after BFP, and 12-24 h after BFP). Piglet BW gain was estimated 24 h after. Data were analysed using proc GLM and MIXED and the probability of sow responsive-ness using proc GENMOD in SAS. Pre-partum nesting 4 h before BFP was related with longer latency of the first sucking ($P<0.01$), with shorter duration of suckling ($P<0.05$) and with lower piglet BW gain ($P<0.05$). More nesting after BFP was associated with a lower probability to react towards the playback of piglets' screams ($P<0.05$). However more postural changes after BFP were associated with a higher probability to react towards the playback of piglets' screams ($P<0.05$). More postural changes during parturition were related with higher piglet BW gain ($P<0.05$), but more postural changes during 12-24 h after BFP were related with lower piglet BW gain ($P<0.01$). No other relationships were detected between-partum nesting and maternal behaviour within the 24 h after BFP. High occurrence of pre-partum nesting 4 h before parturition, but not more pre-partum postural changes, maybe an early indicator for impaired suckling behaviour and lower weight gain during the first 24 h post-partum.

Does behavioural synchrony impact our ability to apply treatments within groups of cows?

Meagan King, Robin Crossley and Trevor Devries
University of Guelph, Animal and Poultry Science, 50 Stone Road East, Guelph, ON, N1G 2W1,
Canada; tdevries@uoguelph.ca

Dairy cows synchronize their behavioural patterns. Studies often have cows on different treatments housed next to each other or assigned to different treatments within a single group. There is potential in such situations for the behaviour of one animal within a group, or housed adjacently, to influence the behaviour of those on another treatment, particularly when treatments may stimulate different behavioural responses through the day. This could minimize the predicted behavioural response to treatments imposed. The objective of this work was to determine if behavioural synchronization has an impact on behavioural response in cows assigned to different treatments within a group. Twenty-four lactating Holstein dairy cows were exposed to each of 2 treatments (over 21-d periods) in a crossover design. Treatments were: (1) delivery of feed at milking time and (2) delivery of feed halfway between milking times. Cows were split into 4 groups, each with 6 cows. Each group of 6 cows was housed in a free-stall pen containing 6 lying stalls, with each cow trained and assigned to consume their ration from an electronic feed trough. In the first 2 groups all cows within a group of 6 were assigned to the same treatment within each period (homogenous treatment group). In the other 2 groups cows were alternately assigned to treatments within each period (mixed treatment group). To test synchrony of behaviour, kappa coefficients were calculated for each animal within each group, as an estimate of agreement that two cows within a group (i.e. each individual and each other cow in their group) would be engaged in feeding activity for any hour of the day. Data were then summarized by cow and treatment and analyzed in a general linear mixed model, specifically contrasting the kappa coefficients for those cows on the same treatment within mixed treatment groups against those on the other treatment in those mixed groups, as well as against those cows in the homogenous treatment groups. Level of synchrony was similar ($P=0.42$) for cows within a homogenous group (kappa=0.29 ± 0.033) compared to those cows on the same treatment within a mixed group (kappa=0.32 ± 0.033). This level of synchrony between cows was nearly 50% greater ($P=0.014$) compared to their synchrony with cows in their group on the other treatment (kappa=0.22 ± 0.033). The results suggest that level of synchronization between cows on the same treatment is similar, regardless of whether there are others in their group on a separate treatment. Further, the level of synchrony with cows on a different treatment in their group is much lower. Thus, synchronization of behaviour does not necessarily restrict our ability to apply different treatments to cows within a group.

The differences in the heart rates of the rider and horse based on riding order for novice and experienced riders

Keiko Furumura and Kyoko Miyanaga
Obihiro University of Agriculture and Veterinary Medicine, Nishi 2-11, Inada, Obihiro, Hokkaido, 080-8555, Japan; kfuru@obihiro.ac.jp

Inexperienced riders have difficulty riding confidently and they tend to feel tension or fear while riding. Generally, when an inexperienced rider follows a trainer or an experienced rider the beginners can handle the horse more confidently with less tension and fear. The purpose of this experiment was to find, (1) if an experienced rider rides first, will the beginner's fear and tension be reduced, (2) if differences in riding order also affected the stress levels of the horse. Four horses were used: two female Thoroughbreds (18 years old), one female Dosanko (18) and one gelding Thoroughbred × Lipizzaner cross (13). A rider with eight years of experience (ER) and five novice riders (NR), with less than 10 rides, were given the same exercise test course. Horse and human RR intervals (horse: S801i & T52H, human: RS800CX & Wearlink®+W.I.N.D., ©Polar, Finland) were measured. The LF/HF ratios were determined using the Polar Pro Trainer5™ software to evaluate the stress level. The average RR interval value and LF/HF ratio for ER/NR ride order and NR/ER ride order, were compared using SAS Enterprise Guide 4.3. On average the RR intervals of the five NRs, were significantally different between the ER/NR ride order and the NR/ER ride order (ER/NR 451±0.9 ms; mean ± SE vs NR/ER 476±0.5 ms, $P<0.0001$). The NRs were divided into two groups. GroupA (three people) was greatly affected by the ride order with a wide fluctuation in the heart rate. GroupB (two people) was not as affected by the ride order and had not noticeable heart rate fluctuations. The LF/HF ratio of GroupA was 503% at NR/ER and 378% at ER/NR. NR tension and stress declined. In the horses, the mean RR interval value is significantly lower with an ER/NR ride order than a NR/ER ride order (ER 830±2.9 ms, NR 906±3.1 ms. $P<0.0001$). It is believed that horse felt more tension with the ER riding. In this experiment, when GroupA NR rode after the ER, the NRs exhibit reduced levels of stress and fear and the horse feels less stress as well. This suggests that ride order should be taken into consideration when training novice riders, so that they can ride more safely.

Relationship between relinquishment reasons and fecal corticosterone level in dogs in a public animal shelter

ChuHan Yang[1], Midori Nakayama[1], Katsuji Uektake[1], Hiromi Uchita[2], Masahiko Akiyama[2], Tsuyoshi Koike[2] and Toshio Tanaka[1]
[1]*Azabu University, School of Veterinary Medicine, 1-17-71 Fuchinobe, Chuo-ku, Sagamihara, 252-5201, Japan,* [2]*Kanagawa Animal Protection Center, 401 Tsutiya, Hiratsuka-shi, 259-1205, Japan; ma1408@azabu-u.ac.jp*

Animal shelters provide a safe and healthy environment for strays and animals relinquished by their owners until introduction into a new home. However, the time each dog spends in a shelter may range from days to years. Besides the duration of stay, dogs are relinquished for various reasons, which are also considered candidate stressors for resident dogs. Thus, how shelters affect admitted animals remains an important issue to be clarified. To determine the stress level, we examined fecal corticosterone concentrations of adult and senior dogs (including medium and large breeds) admitted to the public Kanagawa Animal Protection Center (KAP). The subjects (n=9, 6 males, 3 females) were housed individually in pens (1.0×2.0×2.5 m) with two meals a day and free access to water. Fecal samples were collected from each pen on weekdays from April 21st to August 29th before the morning routine. Feces were frozen at -80 °C in polypropylene bags until assay, and then dried in an oven (100 °C × 16 h) to be powdered. Each 0.1 g powdered fecal sample was added to 1 ml ethanol (95.5%) to extract corticosterone. Finally, the concentration of fecal corticosterone was determined by an enzyme immunoassay kit. Data were analyzed by Student's t-test to compare mean (±SEM) corticosterone concentrations based on relinquishment reason, age, and duration in the shelter. No significant difference was shown between dogs that were relinquished because of an owner factor (e.g. moving, financial, health, aging) or dog behavior (e.g. biting). There was no significant difference between adult (1-5 y) and senior (8-12 y) dogs either. Stays less than 1 month (2,762.0±1,782/9 ng/g) and over 6 months (3,667.0±1,039.8 ng/g) were not significantly different. The concentration at 1-2 months (1,890.0±1,051.5 ng/g) was significantly lower $P<0.05$) than at 6 months.

Evaluation of effectiveness of a pain recognition digital e-learning package by veterinary students in the UK and Japan

Tae Sugano[1] and Fritha Langford[2]

[1]University of Edinburgh, Royal (Dick) School of Veterinary Studies, Easter Bush Campus, Edinburgh, EH25 9RG, United Kingdom, [2]SRUC, Anmal and Veterinary Sciences, West Mains Road, Edinburgh, EH9 3JG, United Kingdom; taebu@aol.com

Pain alleviation is an important part of animal welfare. Veterinary students need to acquire knowledge, recognition and assessment of behavioural and physiological changes associated with pain in animals in order to optimise their future practice. To facilitate this, digital learning objects (LO) on animal pain, including visual and sound aids, was created and distributed online. The LO had the identical versions in English and Japanese languages. Using this, a pilot study on learning effectiveness of the LO and differences in perception of pain in animals among veterinary students by nationality and year of study were tested. Students from the first and final year in Royal (Dick) School of Veterinary Studies (R(D)SVS) at the University of Edinburgh and the second and final year in the School of Veterinary Medicine of Nihon University were chosen as study groups. The students were invited to answer online questionnaires before (a pre-test) and after (a post-test) viewing either the LO or reading material (RM) control in a series of processes. Fifty-nine (33 from 1st year and 26 from 5th year) students took part from R(D)SVS and fourteen (8 from 2nd year and 6 from 6th year) from Nihon. The questions in the pre- and post-tests were related to knowledge, recognition, assessment of and attitude to pain in animals, connected with contents of the LO/RM. The effectiveness of the LO/RM with the UK students was investigated by the change of their responses from pre- to post-test. The responses of pre-test from four study groups were compared in order to look into their differences. Although knowledge and attitudes differed from pre- to post-test, there were no significant differences between LO and RM viewers in knowledge and attitudes post-test. However, the LO seemed to have a positive effect on the students' emotional experience (95% CI, 2.04-2.24; $P=0.06$). In the respect of the students' perception, the differences between the years of study were seen in UK students but not Japanese students. The final year students in R(D)SVS were more likely to express positive attitudes towards pain alleviation than other study groups ($P<0.05$). The results suggest that the students' perception of pain in animals is greatly influenced by what they have learned during the veterinary curriculum. Extra teaching in the form of LOs may encourage students in interaction with the subject matter and its approaches.

Differences in space use fidelity in lame and non-lame dairy cows

Jorge A. Vázquez Diosdado[1], Zoe E. Barker[2], Jonathan R. Amory[2], Darren P. Croft[3], Nick J. Bell[4] and Edward A. Codling[1]

[1]University of Essex, Department of Mathematical Sciences, University of Essex, Wivenhoe Park, Colchester, CO4 3SQ, United Kingdom, [2]Writtle College, Chelmsford Essex, CM1 3RR, United Kingdom, [3]University of Exeter, Center for Research in Animal Behaviour, College of Life and Environmental Sciences, University of Exeter, Exeter EX4 4Q, United Kingdom, [4]Royal Veterinary College, Hawkshead Lane, North Mymms, Hatfield Herts, AL9 7TA, United Kingdom; javazqa@essex.ac.uk

Advances in sensor technology and novel statistical methodologies have made it possible to accurately derive space-use patterns of movement in both wild and domesticated animal species. Until now, space-use fidelity has not been directly linked to differences in the health status of animals. In this study, cows from a high yielding group on a commercial farm were scored for lameness using the DairyCo mobility score. 10 non-lame (score 0) and 10 lame (score 2) cows were subsequently fitted with novel local positioning wireless sensors (Omnisense 500®) mounted on a neck collar. The sensors were used to track the position of each cow within their normal management group in a cubicle shed for 4 days. Space-use fidelity was calculated as the sum of the level of similarity between the space-use between each pair of days for each cow. The space use fidelity values were significantly higher for lame cows (ANOVA: $F_{1,18}=4.45$, $P=0.049$), indicating that lame cows were more likely to occupy the same areas of the barn. Differences in space use fidelity could potentially be used to build a predictive model for early detection of lameness in dairy cows, leading to faster treatment and improved welfare. The methodology could be extended to explore space-use patterns in other animal species.

Measurement of stress-related physiological indexes for a horse with the exercise stress

Hiroshi Yamada and Sae Tsubakisita
Rakuno Gakuen University, Bunkyodai-Midorimachi, 582, Ebetsu, 069-8501, Japan;
yamada@rakuno.ac.jp

A lot of physiological measures have been invented to measure human and animal stress levels, but none of their measures have complete validity. The reactive oxygen metabolites (d-ROMs) and biological antioxidant potential(BAP) are recently focused indexes which are measured with automatic analyzing device. The purpose of this experiment is to confirm what physiological measures, including d-ROMs and BAP are related with exercise stress of a horse. A five years old pony experienced two sessions of 15 minutes exercise. In the exercise, the horse were tied with a horse training rope and run at a trot. The same exercise procedure were employed in two sessions. In each session, blood samples were taken five times, prior to, 1, 5, 10, and 15 minutes after the exercise. Heart rates were also taken with Polar Heart Rate sensor continuously throughout the experimental session. Lectal temperature, levels of lactic, cortisol, d-ROMs, BAP, chromogranin A (CgA), counts of lymphocyte, neutropenia, and white blood were quantified with each blood sample. Cortisol level, which generally increases with the experience of temporal stressor, raised 1 or 5 minutes after the exercise, then maintained until 15 minutes (from 4.0 µg/ml to 4.0, 4.6, 5.4 and 5.2). For BAP, a reports shows an increases after the strong exercise for human subjects, increased BAP level were observed 5, and 10 minutes after the exercise in this experiment (from 2,563.4 µmol/l to 2,510.8, 2,568.2, 2,586.3 and 2,457.2). White blood counts raised up rapidly 1 minutes after the exercise, then declined gradually (from 5,575 counts/µl to 7,250, 6,600, 6,400, and 6,400). Lectal temperature (from 37.8 °C to 38.4, 38.4, 38.4, and 38.2) and heart rates also raised up (from 40.0 bpm to 69.3, 51.5, 47.7 and 47.3) after the exercise. Other measures, CgA and the counts of neutropenia, did not change systematically. The white blood count indicate the exercise in this experiment produces physical stress to the horse, and the conclusion is supported with the level of cortisol and BAP. The time delay between the stressful event and the appearance of the stress responses should be cared as shown in cortisol level. In this experiment, again, highest values in cortisol and BAP were not observed 1 minutes after the exercise, that would be most stressful period of all measurements. Although this is a case study with a single subject and lacks the reliability, this data informs the relationship between the exercise stress and newly developed physiological indexes.

Effects of light intensity on rhythmicity of core body temperature and body lesion scores in growing pigs

Jialan Zheng, Stephan Schäfer, Stephan Ebschke, Christian Nawroth and Eberhard Von Borell
Institute of Agricultural and Nutritional Sciences; Martin-Luther-University Halle-Wittenberg, Animal Husbandry and Ecology, Theodor-Lieser-Strasse 11, 06120 Halle, Germany; eberhard.vonborell@landw.uni-halle.de

Little is known about the effects of light intensity on the rhythmicity and behavioural activity of growing pigs. The anatomy and physiology of the porcine eye suggest that domestic pigs are best adapted to dim levels of natural full-spectrum light. On the other hand, behavioural problems such as tail biting and aggression are often attributed to light conditions in artificially illuminated animal facilities. Our goal was to study effects of full-spectrum light intensities on rhythm of core body temperature (CBT) and skin lesion scores (SLS) under laboratory conditions (temperature: 23±1 °C, humidity: 65±5%). Four groups of 4 pigs in each of 2 treatments from 8 to 12 wks of age were used. Pigs either started with a period of 2 wks (Period 1) with an illumination of 600 lux followed by 2 wks (Period 2) of 80 lux [HL], or underwent treatment in a reversed order [LH]. For both treatments, full-spectrum illumination lasted 8 hours (from 9:00 to 17:00 h) for all pigs with a red light of 2 lux in the scotophase. We utilized CBT measured in 10 min intervals as an indication for rhythmicity, and weekly SLS (0-6) from 16 body locations as an indirect measure of aggression level within each group. Linear mixed models (REML) and ANOVA were used for analysis (SPSS). Independent of treatments, CBT followed a biphasic pattern as this is also reported for behavioural activities with maximum peak temperatures between 10:00 to 11:00 and 15:00 to 17:00 h. CBT was in general slightly increased in the HL treatment group, independent from light intensity in total (HL 40.49±0.07 vs LH 40.29±0.07 °C; $F_{(1,27)}=4.68$, $P=0.03$) and within periods (Period 1: HL 40.52±0.05 vs LH 40.32±0.08 °C; $F_{(1,27)}=4.90$, $P=0.04$; Period 2: HL 40.46±0.06 vs LH 40.27±0.08 °C; $F_{(1,27)}=3.82$, $P=0.06$). The aggregated SLS were in general higher in the HL groups (HL 19.67±0.74 vs LH 13.87±0.76; $F_{(1,29)}=30.06$, $P<0.00$ 1) and were increased under periods of 600 lux (Period 1: HL 20.25±0.66 vs LH 8.80±1.00; $F_{(1,29)}=93.18$, $P<0.001$; Period 2: HL 19.09±0.78 vs LH 18.93±1.04; $F_{(1,29)}=0.02$, $P=0.89$), although scores in the HL groups remained high under 80 lux once they were established in Period 1. The results indicate that a daylight intensity of 600 lux under laboratory conditions affected behavioural activities that are associated with increased SLS and that CBT pattern followed independently from light intensity the same biphasic diurnal activity rhythm as reported in previous behavioural studies with pigs.

The effects of anti-motion sickness medication on the responses to road transportation in goats

Masato Aoyama, Kotomi Sanuka, Yoko Takakusaki, Takumi Motegi, Hiroki Kaneta and Shoei Sugita
Utsunomiya University, Department of Animal Science, 350 Minemachi, Utsunomiya-city, 321-8505, Japan; aoyamam@cc.utsunomiya-u.ac.jp

It is unclear whether ruminants experience motion sickness during road transportation, because ruminants do not vomit. This study examined the possibility of the motion sickness as one of the factor of the transportation stress in domestic ruminants. In experiment (Exp) 1, we examined the effects of diphenhydramine (DH), which is one of the anti-motion sickness agents for humans, on behaviors during road transportation in goats. In Exp 2, we examined the effects of road transportation on the c-Fos protein (one of the marker substance indicating the activation of the neurons) expression in the caprine brains stem, which is involved in the induction of the motion sickness in other animals. In Exp 3, we examined the effects of DH administration on the c-Fos expression in the caprine brainstem induced by the transportation. In Exp 1, eleven adult Shiba goats were used. They were intramuscularly (IM) administrated with DH 15 minutes before the start of the transportation, and the behavior of each goat was observed. Each goat was administrated with the same volume of saline and transported on another day for the control. Some goats showed the specific postural shapes that faced down and little moved even when we approached them during the transportation, but its duration in DH administrated goats was significantly shorter than those of saline administrated goats ($P<0.05$, Wilcoxon's matched-pair signed-rank test). In addition, the frequency of the attempts to escape from our approach in DH session was tended to be higher than those in the controls ($P=0.068$). In Exp 2, four adult Shiba goats were used. Two goats were transported for 1 hour, whereas other two goats were loaded on a parked truck as the control. The c-Fos expressions in the brain stem were visualized by immunohistochemistry, and the numbers of the immunoreactive cells were counted. The numbers of the c-Fos protein immunoreactive cells in the nucleus tract solitary (NTS), which is involved in the induction of nausea and emesis in other animals, were higher in the transported goats than in the controls. In Exp 3, six goats were used. Three goats were IM administrated with DH 15 min before the transportation onset whereas other three goats were administrated with saline as the control. Although it was not statistically significant, the numbers of the c-Fos immunoreactive cells in the NTS in DH administrated goats were lesser than those in saline administrated goats. These results indicate that goats might experience motion sickness derived from road transportation.

The behavior of White Leghorn hens with trimmed comb and wattles before, during, and after a heating episode

Patricia Y. Hester[1], Diya S. Al-Ramamneh[2] and Maja M. Makagon[1]
[1]Purdue University, Department of Animal Sciences, 125 S Russell St., West Lafayette, IN 47907-2042, USA, [2]Jerash Private University, Department of Animal Science and Food Nutrition, P.O. Box 311, Jerash 26150, Jordan; phester@purdue.edu

Nineteen million pullets annually in the United States are subjected to a partial trim of the comb, but not the wattles, to improve feed efficiency. During high temperatures, the superficial blood vessels of the comb and wattles vasodilate allowing blood to pool to the skin's surface resulting in conductive cooling. The objective of this study was to determine the effect of a partial comb and wattle trim on the ability of Leghorn hens to perform thermoregulatory behaviors before, during, and after an imposed heating episode. At 21 days of age, the comb and wattles of 78 pullets were trimmed using surgical scissors that were disinfected with alcohol between trims as described in the 2010 Ag Guide, with a additional 78 pullets serving as controls. Standard farming protocol for trimming was followed in which analgesic was not used. At 189 to 191 days of age, hens were subjected to a heating episode (34.6 °C for 50.5 h). Live observations of behavior were conducted 3 times daily from 08:00 to 09:00, 12:00 to 13:00, and 15:00 to 16:00 h before, during, and after the heating episode using instantaneous scan sampling at 5 min intervals. The numbers of hens in each cage that were drinking, feeding, standing, sitting, panting, and spreading their wings were recorded. Feed consumption was determined at 25, 30, and 35 wk of age. Data were subjected to an ANOVA. The heating episode caused an increased proportion of trimmed hens to pant (41 vs 28% for trimmed and control hens, respectively, SEM=2, P=0.001). All other measured behaviors were not affected by the trimming treatment. During an exceptionally hot day later in the summer when room temperature naturally peaked at 31.3 °C at 1514 h, a greater proportion of the trimmed hens panted (55 vs 15% for trimmed and control hens, respectively, SEM=4, P<0.0001) and spread their wings (54 vs 30% for trimmed and control hens, respectively, SEM=4, P=0.005) from 15:00 to 16:00 h. Less feed was used (92.3 and 98.1 g for trimmed and control hens, respectively, SEM=1.2, P=0.001) in trimmed hens as compared to controls with no effect on feed efficiency (P=051). The summer heat may have been a major factor in contributing to the lower feed usage of trimmed hens because of their inability to thermoregulate as effectively as non-trimmed controls. In conclusion, the increased proportions of trimmed hens panting and spreading their wings suggest that hens need their combs and wattles to thermoregulate effectively during periods of high environmental temperature.

Applying ethology to the development of international farm animal welfare standards and recommendations

Sara Shields

Humane Society International, Farm Animals, 2100 L Street, N.W., Washington, DC 20037, USA; sshields@hsi.org

A number of new global animal welfare standards and recommendations are currently being developed. Ethologists can – and do – make a valuable contribution to these standards, but greater input is needed. The World Organization for Animal Health (OIE) publishes the Terrestrial Animal Health Code, which establishes guidance on animal health and zoonosis in international trade. Since 2001, when animal welfare was identified as a priority in the OIE Strategic Plan, chapters on animal transport, slaughter, and farm management have also been adopted. In 2012, the International Organization for Standardization (ISO) began working on a Technical Specification (TS) to serve as a tool in business-to-business relationships for implementing OIE animal welfare chapters as well as private standards. The ISO working group on animal welfare expects to complete the TS in 2016. The Responsible Down Standard (RDS) and the Responsible Wool Standard (RWS) are both projects of Textile Exchange, a non-profit sustainable textiles organization. The RDS, which covers waterfowl used for down production, was launched in 2014, and underwent revision in 2015. The RWS is currently being drafted by an international, multi-stakeholder group. The RDS and the RWS are global, auditable certification schemes. Both the World Bank's International Finance Corporation and the European Bank for Reconstruction and Development have animal welfare requirements for their agribusiness investments. Ethology is essential to the development of these and other science-based standards. Ethological research has identified behavioral needs, has been used to establish animal-based measures, can aid in identifying painful procedures, and can help delineate the features necessary for comfortable living environments, to name just a few examples. While veterinarians play a key role in making sure animal health issues are covered in international standards, the role that ethologists play in ensuring animal welfare is often less well recognized. Ethologists may want to consider becoming more involved.

An impact assessment tool for an equine welfare project in India

Azeem Ahamad and Dinesh Mohite

Brooke Hospital For Animals, Welfare Assessment, 2nd Floor, Pacific Business Park, Dr. Burman Marg Plot no 37/1, Site IV, Sahibabad Industrial Area, 201010, Ghaziabad, Uttar Pradesh, India; azeem@thebrookeindia.org

The Brooke is an international equine welfare organization working in eleven countries in Asia, Africa and America. Brooke India is an affiliate of The Brooke and its core value is sustainable equine welfare through human behavior change. Change was achieved through community meetings, training of stakeholders and formation of equine welfare groups to undertake collective work to reduce welfare issues. Aim of the study was to assess the effectiveness of standardized equine-based welfare assessment tool (SEBWAT) in documenting equine welfare changes. SEBWAT comprises 40 equine health and behavioral indicators. SEBWAT was used to identify level of welfare in equine after two years intervention activities of JP Nagar district comprising owner-equid compassion building, grooming and harness care, improved feeding and hoof care. In first assessment (2011-12), 32 of 60 villages were randomly selected from JP Nagar 372 equids (horses=329, mules=39 and donkeys=4) were assessed. In second assessment (2013-14), 27 of 32 (selected during 2011-12) villages were selected and 326 equids (horses=303, mules=22 and donkeys=1) were assessed. Then data were entered, analyzed and compared using customized software. Reductions in apathetic behaviour of equids (5% to 1%, $P=0.002$), ectoparasite infestation (46 to 31%, $P<0.001$), evidence of diarrhoea (18 to 8%, $P<0.001$) and severe body wounds (29 to 25%, $P=0.27$),with white mucous membranes to pink (15 to 3%, $P<0.001$), abnormal hoof shape (65 to 47%, $P<0.001$), poor horn quality (41 to 21%, $P<0.001$) and poor frog health (29 to 11%, $P<0.001$) were recorded. Percentage s of reductions was different as per different issues. Intervention done by JP Nagar has shown improvements in equine. SEBWAT was successful in identifying quantum of improvement in equine welfare as well as areas lacking behind. It is concluded that SEBWAT may be an effective tool for programme evaluation which facilitates organizational learning and provides scientific evidence that can underpin reflection on design of interventions.

Evaluation of heart rate variability of grazing cows by simultaneous measurements of RR intervals and body acceleration

Kazato Oishi[1], Masafumi Miwa[1], Yudai Yasunaka[1], Sawako Yamada[1], Hideyo Kitaura[2], Hisaya Sakai[2], Kanji Yamaguchi[2], Yukiko Himeno[3], Hajime Kumagai[1], Seiji Ieiri[2] and Hiroyuki Hirooka[1]
[1]Kyoto University, Graduate School of Agriculture, Kitashirakawa-Oiwake-Cho, Sakyo-ku, Kyoto 6068502, Japan, [2]Kumamoto Prefectural Research Center, Grassland and Animal Husbandry Division, 1454 Nishiuura, Aso, Kumamoto 8692304, Japan, [3]Ritsumeikan University, Department of Bioinformatics, College of Life Sciences, 1-1-1 Noji-higasi, Kusatsu, Shiga 525-5577, Japan; kazato@kais.kyoto-u.ac.jp

Measurement of heart rate variability (HRV) is a non-invasive method that can be used to monitor the autonomic nervous system activity. Monitoring the autonomic nervous system activity in farm animals has recently gained interest as a non-invasive technique to assess animal welfare, because the vagal component of the autonomic nervous system plays an important role in regulating heart rate (HR) in response to stress. However, the HRV measurements can be greatly affected by behavioral loads of animals. On the other hand, an acceleration index, overall dynamic body acceleration (ODBA), has recently been proposed in the field of wild animal ecology to quantify the activity level of animals with accelerometers. The objective of the present study was to examine the possibility of using ODBA together with the analysis of HRV to assess the autonomic nervous system of grazing cows. Three Japanese Brown cows were equipped with two separate electrodes of electrocardiogram for the HR monitor and an acceleration logger, and the RR intervals and body acceleration were simultaneously measured under grazing and housing managements. The recorded RR intervals were used to calculate HRV indicators in three analyses (time-domain analysis, spectral analysis and nonlinear analysis), and the effects of the difference in managements (grazing vs housing) and activity levels expressed by ODBA on the indicators were analyzed. The results showed that most of the HRV indicators were strongly affected by ODBA, and that the effect of the managements on the indicators in spectral and nonlinear analyses could be large with the increase in ODBA. At high ODBA, the cows under grazing showed significantly higher high-frequency components and lower short-term detrended fluctuation components.

Monitoring behavior in cows using a three-dimensional accelerometer and ZigBee network

Keiji Takahashi[1] and Hidetoshi Watanabe[2]
[1]Rakuno Gakuen University, Department of Sustainable Agriculture, 582 Bunkyodai-Midorimachi, Ebetu, 069-8501, Hokkaido, Japan, [2]Orion Machinery Co., Ltd., 246 Kotaka, Suzaka, 382-8502, Nagano, Japan; takahakj@rakuno.ac.jp

At dairy facilities, monitoring of cows' behavior is useful for evaluating animal comfort. As it is not feasible for individuals to conduct 24-h observation, in this study a behavior-monitoring sensor for cows was developed using an accelerometer and ZigBee network. Cow behaviors, such as lying, standing, walking, and eating, were detected with a three-axis micromachined accelerometer attached to one of the front legs. Three-dimensional acceleration data were converted from the native analog format to digital using an A/D converter. Acceleration data were stored using 512-kbit EEPROM and sent to the processing unit via I^2C. ZigBee modules were used to create a wireless mesh network, and a ZigBee module with one Coordinator API and six Router ATs was used as the end device. Acceleration data were measured every 0.1 s and transmitted through the ZigBee network at 12- or 120-s intervals. The programs used in the processor are mostly open source and the PC's data-acquisition program was programmed using Microsoft Visual Studio 2013. Three AA batteries were used, for a total battery life of 5-6 days. The price per sensor was approximately 6,500 JPY (US$55) and the price for the receiver unit was approximately 7,500 JPY (US$65). With this system, the behavior of 6 cows in a freestall barn was recorded for 2-4 h and their behavior was also visually monitored. Acceleration along the x, y, and z axes was analyzed, and the data clearly indicated whether cows were standing, walking, or lying. However, when a cow folded its front leg under the body, acceleration data could not be received and the program stopped recording. Minor improvements were made, after which the behavior of 12 cows in a tie-stall barn was recorded for 5-6 days. As before, the data could sometimes not be received when a cow folded its leg under its body, but when the cow stood, data transmission quickly resumed. This method is therefore useful for monitoring the behavior of individual cows over a period of several days, without the need for tedious 24-hour visual monitoring. Monitoring whether a cow is lying or standing is useful to infer the condition of a cow's legs and to detect the cow's heat in the tie-stall barn. Such monitoring is useful to evaluate whether the stalls in freestall and tie-stall barns are comfortable. This three-dimensional acceleration behavior-monitoring system is currently a laboratory-built system, but further development is planned toward commercialization and larger scale manufacture.

Grooming device use and effects on self and social grooming behaviour in fattening beef cattle

Shigeru Ninomiya, Kie Higashi and Yoshito Moriyasu
Gifu University, Faculty of Applied Biological Sciences, 1-1 Yanagido, Gifu, 5011193, Japan;
nino38@gifu-u.ac.jp

Environmental enrichment has been conducted to stimulate animal behaviour in captive housing. Scientific approaches to environmental enrichment must investigate animals' use of enrichment devices and assess enrichment effects on the animals. For this study, Japanese Black fattening cattle were provided with a brush as an enrichment device. We then observed their brush usage and investigated effects of its introduction on their self-grooming and social grooming behaviour. Experiment 1 examined three Japanese Black steers that were 23.7±0.6 months old. They were singly housed in a 3.0×2.8 m pen. Experiment 2 examined another six Japanese Black steers that were 19.3±4.6 months old, paired in the same pens. The animals were given free access to water and followed the same feeding regime. One 50-cm-long and 10-cm-wide brush was arranged with a second brush in each pen in an inverted L-shaped pattern. The self and social grooming behaviour and brush usage were recorded continuously for 48 h in Experiment 1 and for 30 h in Experiment 2 before and about one month after introducing the brush. The behaviour durations were analysed using GLM. The steers used the brush for rubbing the head (43.5% of total use), tail (15.6%) neck (10.0%), back (8.8%), and hip cross (4.6%); 15.8% of the total use was licking it. By introducing the brush, self-grooming (excluding brush usage) tended to decrease in Experiment 1 ($P<0.10$) and decreased in Experiment 2 ($P<0.05$). Social grooming was decreased by introducing the brush in Experiment 1 ($P<0.05$), but not in Experiment 2. These results suggest that the steers use the brush instead of self-grooming. Introducing a brush can apparently decrease social grooming by fattening beef cattle according to housing conditions.

Mechanical brush usage by female Holstein calves separated from their dams immediately after birth

Ken-ichi Yayou[1] and Mitsuyoshi Ishida[2]
[1]National Institute of Agrobiological Sciences, Tsukuba, 305-8602, Japan, [2]Institute of Livestock and Grassland Science, Tsukuba, 305-0901, Japan; ken318@affrc.go.jp

Although early cow-calf separation has several benefits for farmers and animals, early maternal deprivation has long lasting negative effects on the growth and behavioural development of several animal species other than cattle. We hypothesized that providing early-separated calves with an alternative for their mother's care could improve their welfare and also have a favorable impact on their growth and behavioural development. Among the several components of maternal care, we focused on mother's grooming and developed an automatically rotating mechanical brush as an alternative to simulate maternal grooming of calves and examined its effects on the growth and behaviour of female Holstein calves. The calves were raised in individual pens without mechanical brushes (Cont; n=5) or with mechanical brushes (Bru; n=5) immediately after birth. At approximately 45 days of age, they were weaned. At approximately 2 months of age, they were introduced into a group of 3-4 older calves. The Bru calves utilized the brush consistently until grouping. The average utilization time/day until weaning was 15.3±12.9 (SD) min. The usage frequency was 26.2±11.6 (SD) times/day. The average usage time/episode was 36.8±26.6 (SD) s. The brush was used more frequently for neck grooming than for belly grooming at 1 month of age (Scheffe test, $P<0.1$). The average daily weight gains until weaning, 0-3 months after grouping, and 10 months after birth were similar between the groups. The time budget (standing or lying) and self-grooming total episodes during 24 hours at 1 month of age were similar between the groups. At 1 month of age, the latency to first exploration of a novel object introduced into their home pen was significantly shorter in the Bru than in the Cont calves ($P<0.05$; Mann-Whitney U test). The total time in contact with the novel object was longer in the Bru calves ($P<0.05$). The present study suggested a strong motivation of calves toward the use of a mechanical brush developed as a substitute device for maternal grooming. Although further research is needed, one of the possible positive effects of the mechanical brush was to reduce fear and fearfulness for the novel object introduced into the home environments of female calves.

Tethering period effects on cattle jumping behaviour when released to the paddock

Fuko Nakayama and Shigeru Ninomiya
Gifu University, Applied Biological Science, 1-1 Yanagido Gifu, 501-1193,Gifu, Japan;
u8122027@edu.gifu-u.ac.jp

Tethering is expected to be stressful for cattle because they are unable to express appropriate behaviour when motivated. But it remains unclear. This study was conducted to investigate whether jumping of tethered cattle released indicates motivation during tethering. We examined 16 Japanese black cattle of 49.5±21.5 months mean age. They were divided into four groups. The animals were tethered with a rope in an indoor pen and released to a neighboring outdoor paddock every day: 9:30-14:30. Subsequently by preventing their release for one day or 6 days, we observed responses to being tethered for three durations (19 h, 43 h, and 163 h). After each tethering period had finished, animals were released to the paddock for 5 h. Then their behaviours were video-recorded for 60 min immediately after release. From the vido, the frequency of jumping was observed. Jumping behaviour was defined as the status of four legs, forelegs or hind legs lifted from the ground. Additionally, the running (trotting and galloping) frequency was observed. To investigate which behaviour was restricted by tethering, we filmed with a CCD camera of the animals tethered from the second day to the third day during 163 h tethering treatment. Various behaviours were compared with the behaviours exhibited when individuals were released after 19 h tethering. The frequencies of jumping per individuals per hour were 1.25±3.47 in 19 h, 3.38±5.34 in 43 h, and 7.44±12.21 in 163 h (mean ± SD). Jumping occurred more after 163 h than after 19 h ($P=0.016$). The frequencies of running was 0.88±1.09 in 19 h, 1.75±1.84 in 43 h, and 3.25±3.68 in 163 h. Running was observed more after 163 h and 43 h then after 19 h ($P=0.005$, $P=0.005$). When cattle were tethered, jumping, running and sexual behaviours were not observed. These results suggested that when cattle were tethered for 6 days, jumping and running after being released to the paddock reflect the motivation built up by behaviour restriction during their tethered periods.

Effect of a landscape immersion exhibit on the behavior of captive white-handed gibbons (*Hylobates lar*)

Y. Kimura[1], K. Takashi[1], M. Tamechika[1], A. Yatsushiro[2] and S. Ito[3]
[1]Tokiwa Zoo, 3-4-1, Norisada, Ube-shi, Yamaguchiken, 755-0003, Japan, [2]Tokai University, Graduate school of Agriculture, Kawayo Minamiaso-mura, Aso-gun, Kumamotoken, 869-1404, Japan, [3]Tokai University, School of Agriculture, Kawayo Minamiaso-mura, Aso-gun, Kumamotoken, 869-1404, Japan; kimura@tokiwa-zoo.jp

White-handed gibbons are listed as an endangered species on the International Union for Conservation of Nature (IUCN) Red List. In Japan, this species is popular in zoos. In March 2015, at Tokiwa Zoological Garden, the white-handed gibbon exhibit was changed from an old cage-style to a landscape immersion exhibit. Landscape immersion has vegetation and natural landscape components in outside exhibition facilities and is intended to give visitors the sense that they are actually in the animals' habitat. There are few investigations about the effects of landscape immersion on captive animal behavior. We observed the behavior of the white-handed gibbons to investigate behavioral changes when they were moved from the old cage-style housing (concrete) to a landscape immersion exhibit (with plants and other natural vegetation). From September 2014 to January 2015, white-handed gibbons (n=5) were observed by the one min scan-sampling method for 16 behavioral categories in both exhibition facilities. The behavioral data were analyzed using the paired t-test. The proportion of time spent on foraging and drinking was significantly greater ($P=0.0146$) in the landscape immersion ($7.8\pm3.5\%$) than the cage-style housing ($2.3\pm0.8\%$). Due to plant consumption in the landscape immersion exhibit, gibbons spent more time eating in that exhibit than in their former housing. The proportion of time spent resting was significantly less ($P=0.0218$) in the landscape immersion ($28.2\pm15.0\%$) than the cage-style housing ($54.5\pm30.2\%$). Gibbons spent more time in the indoor area of the exhibit when housed in the landscape immersion exhibit ($37.2\pm14.1\%$). The entrance to the indoor area was always open and the outside temperature was low (average temperature 6.4 °C), which may explain the reduced rest time and greater time spent indoors. Locomotory behavior, including brachiation, was not significantly different in the landscape immersion style ($15.5\pm8.1\%$) when compared with the cage style ($19.1\pm9.2\%$). There were no significant differences in other measured behaviors between the two exhibits. These results suggest that landscape immersion exhibits have minor effects on the behavior of captive white-handed gibbons.

Effect of cow brush by temperature difference on the growth performance and lying behavior of the Hanwoo steers

Ka Young Yang[1], Kang Hoon Lee[1], Gie Won Kim[1], Joo Hun Kim[1], Yeon Soo Park[2], Chang Woo Lee[2], Jae Jung Ha[3], Seung Hawn Sin[1] and Young Han Song[1]
[1]Kangwon National University, College of Animal Life Sciences, KNU Animal Welfare Management Research Lab. Animal Life Science Bldg 1-204, 200-701,Chun-cheon, Korea, South, [2]Gangwon-do Livestock Research Institute, 1028-3 Hyeancheon li, Doonae meon, 255-831, Hwangsung gun, Korea, North, [3]3Gyeongsangbuk-do Livestock Research Institute, mook li, 750-871, Youngjoo-si, Korea, South; yang.kayoung@gmail.com

This study was conducted to investigate the growth performance and lying behavior of fattening steers. Twenty-eight steers (7 heads/pen, aged 24-months-old) with an on average body weight of 619.8 kg in fattening period from a period of six months. Statistical analysis was carried out using the Mixed Procedure of SAS. For the control group, a cow brush was not implemented, while a cow brush was installed at the center of the right side of the pen for the treatment group. The average daily gain was measured in a two-month interval after the beginning of the experiment. In order to observe the lying, feeding, standing, walking, brushing, and mounting behaviors of individual group. The behavior of the Hanwoo steers was observed from the shoot images for 12 hours during the day (06:00 am – 6:00 pm) from the four installed CCTV cameras (IR LED Camera; APD-7070V), with the behavioral changes noted every two minutes. The behavioral characteristics of the steers were observed 2-4 in a month according to the maximum temperature and minimum temperature. Temperature measurement was carried out with Testo thermometers (174T- 1 channel mini-temperature locker) that were installed at the entry point and exit point of the pen, as well as at two meters above the center of the pen. The lying behavior according to the temperature difference was not significantly different ($P>0.05$) during the period of maximum temperature (average of 26.1 °C); the same goes for the feeding, standing, and walking behaviors between the groups ($P>0.05$). Fighting was observed more in the control group than in the treatment group with a significant difference ($P<0.01$). Further, the brushing behavior was significantly higher in the treatment group ($P<0.01$) under the maximum temperature condition. Under the minimum temperature condition (average of 13.9 °C), the lying behavior was significantly higher in the treatment group than in the control group ($P<0.01$). The installation of a cow brush could give a positive effect on the relieving the stress from the steers by relieving from the itching and by removing foreign materials from the fur.

Suspended pieces of wood as environmental enrichment for commercially housed pigs: effects on animal-based welfare measures

Paolo Cornale[1], Elisabetta Macchi[2], Silvia Miretti[2], Manuela Renna[1], Carola Lussiana[1], Giovanni Perona[2], Luca Battaglini[1] and Antonio Mimosi[1]
[1]University of Torino, Agricultural, Forest and Food Sciences, Largo Paolo Braccini, 2, 10095 Grugliasco (TO), Italy, [2]University of Torino, Veterinary Science, Largo Paolo Braccini, 2, 10095 Grugliasco (TO), Italy; paolo.cornale@unito.it

Animal-based measures are part of the Welfare Quality® protocol as effective indicators to assess pigs' welfare at farm. The present study aimed to determine if the introduction of suspended pieces of wood as environmental enrichment may affect animal-based welfare measures of growing-finishing pigs under commercial farm conditions. In a commercial pig unit, 513 growing pigs were housed indoors into 27 pens (19 animals/pen, 1.0 m²/pig). While half of the pens remained barren, the other half was equipped with environmental enrichments: two pieces of wood suspended on a chain at pig head level. Twenty pens (10 barren+10 enriched) were randomly selected and assessed at 2-week intervals during the trial (20 weeks). Manure on the body, wounds, bursitis, and lameness were scored at individual level according to a 3-point scale; tail biting and hernias were scored on a 2-point scale. Furthermore, the number of pigs removed from the experiment due to health problems and the causes of removal were monitored. Prevalence of each measurement was expressed as percentage of individuals affected on the total of assessed animals. Data were statistically treated using a generalized linear model, with enrichment as fixed factor. The number of pigs removed from the experiment did not statistically differ between treatments. The main causes of pigs' removal were tail biting and lameness. The pigs housed in the enriched pens showed a lower prevalence of wounds (score 1) than barren-housed pigs (3.87 vs 5.43%; $P \leq 0.05$). On the contrary, the pigs kept in the enriched condition showed higher (although not significant) prevalence of lameness (score 1: 1.97 vs 1.07%) and bursitis (score 1: 3.29 vs 2.14). The others indicators did not differ between treatments. Even though the higher lameness prevalence detected in the enriched-housed pigs requires further investigations, the reduced incidence of moderate wounds suggests that suspended pieces of wood might be used as effective environmental enrichments for pigs.

Flexible and adaptable changes in behavior of grey kangaroos (*Macropus giganteus*) kept in a zoo in Japan

Katsuji Uetake[1], Yuko Wada[2], Wen Li Wen[1] and Toshio Tanaka[1]
[1]Azabu University, Sagamihara, 252-5201, Japan, [2]Kanazawa Zoo, Yokohama, 236-0042, Japan; ma1505@azabu-u.ac.jp

Captive animals kept in intensive systems like zoo cannot express their behavioral repertoire as fully as they do in natural environments. However, grey kangaroos (*Macropus giganteus*) show high reproductive success in zoos even though they cannot behave as they do in natural envronment. We compared the behavior of grey kangaroos (n=29) kept in a Japanese zoo with behavior reported for this species in less restricted conditions. The behavior of subject animals was observed continuously over three days, totaling 72 hours of observation, using a scan-sampling method at 20-min intervals. During daylight hours, observation was carried out in an outdoor paddock (400 m^2) that had a soil substrate, scattered rocks and trees, and a watering place. Evening and night behaviors (16:20-8:40 h) were observed in an indoor room (20 m^2) with the aid of two infrared radiation recording devices. Timothy hay (15 kg) and pelleted feed (5.5 kg) were delivered to the indoor room in the evening. Approximately 2 kg of fresh sorgo grass was provided in the morning and afternoon in the outdoor paddock. Z-tests in 'Statcel 3' were used to compare mean (± SEM) daily behaviors observed in the zoo with those reported in a semi-natural environment. Zoo kangaroos spent less time foraging (25.0±2.5%; Z=-68.5) and moving (1.8±0.3%; Z=-328.3), while they spent more time resting (63.4±2.3%; Z=60.6) and in other behaviors such as exploring and grooming (9.9±0.0%; Z=106.2) (all $P<0.001$). In addition, grey kangaroos flexibly changed their nocturnal or crepuscular behavioral patterns to diurnal ones in response to the daily management routine at the zoo. The results indicate that grey kangaroos exhibit a high degree of behavioral plasticity, facilitating their adaption to an intensive zoo system.

The effects of environmental complexity on fear responses of broiler chickens

Peta Simone Taylor, Paul Hamilton Hemsworth and Jean-Loup Rault

The University of Melbourne, Animal Welfare Science Centre, Monash Road, Parkville, 3010, Australia; petat@student.unimelb.edu.au

Environments lacking complexity can decrease exploratory behaviour and adaptation to fearful experiences. We investigated the effects of increased environment complexity on fear responses of broiler chickens. Day-old mixed-sex ROSS 308 broilers (n=108) were randomly allocated to either a control (C), physical environmental complexity (P) or human contact (H) group. Birds were housed in nine groups of 12 in 1 m^2 pens, visually isolated from other pens. Control pens mimicked commercial settings; wood shaving litter, a drinker and feeder and only routine human contact (<10 min/week). Physical items for P birds included a perch, rope, chains, balls, shredded paper and a coloured drawing. The H birds were exposed twice daily to a human walking through the pen for five min with intermittent stationary periods in addition to routine contact. Individual birds were tested at either 21 or 35 days of age (n=6/pen/age) for their fear responses in tonic immobility (TI), novel object (NO) and open field (OF) tests over three consecutive days. A human approach (HAP) and avoidance (HAV) test was performed on day 38. Data were analysed by non-parametric Kruskal-Wallis tests. Duration of TI did not differ between C, P or H birds at either age; however P birds tended to need a greater number of attempts to induce TI at 21 days (C, mean 1.56 ($\pm$ 0.17); P, 2.33 (0.29); H, 1.64 (0.21); $P \leq 0.06$). No differences were seen between C, H or P birds at either age in the NO test. The H birds remained closer to the human in the HAP (C, mean 65.55 ($\pm$ 3.44); P, 65.54 (3.21); H, 54.82 (3.34); $P<0.05$). In addition both H and P birds had a shorter flight distance to the human upon approach during the HAV test than the C birds (C, mean 40.16 ($\pm$ 8.44), P, 30.72 (8.16), H, 27.63 (7.54); $P<0.05$). Although there appears to be some effect of additional physical items on fear responses by day 21, the evidence is not clear nor do the effects appear sustained. Physical enrichment may increase adaptation to potential fearful environments such as range access (permitted at day 21 in Australia), although the minimal effects on fear response observed in this study may prove to be insufficient. This study also showed a decreased fear response specific to humans when birds were raised in a complex environment, either with physical items or human contact. Such a response may alter bird adaptation to potential fearful activities i.e. transportation or routine husbandry duties. These results highlight the potential of increasing environmental complexity to decrease fear responses in broilers; however such effects may be strongly related to the type of items provided or novelty.

The invisible presence: variability in olfactory profiles of husbandry, and impact on laboratory mouse welfare

Noelia Lopez Salesansky, Charlotte C Burn, Nur Hidayu Mazlan, Dominic J Wells and Lucy Whitfield
Royal Veterinary College, Hawskhead Lane, AL9 7TA Herts, United Kingdom; nlopez@rvc.ac.uk

Mice are influenced by scents from other animals and the environment that can cause stress and defensive responses, potentially confounding experimental research and affecting mouse welfare. Husbandry procedures require personal protective equipment and sanitizing products (cleaning agents and gloves) that can be potentially odorous and aversive for mice. A survey completed by 61 individuals working in 52 different UK research institutions, revealed that most (70%) respondents always wore gloves for handling mice; 30% sometimes used bare hands. Nitrile was the most common glove material (94%), followed by Latex (23%) and Vinyl (14%). Between 7 and 19% of the respondents thought that cleaning products definitely, or were likely to, have strong effects on standardization, mouse health, physiology or behaviour. Trigene™ (now called Anistel™) was the most common product used for most cleaning practices. A later survey of 30 people revealed differences in the olfactory perception of glove materials, with latex being significantly more odorous than nitrile. To determine how glove material or bare hands affected mouse behaviour an experiment was performed using C57BL/6J and DBA/6J mice of both sexes (12 same-sex pairs). In a repeated measures design, mice were handled with latex or nitrile gloves or bare hands, and behaviours including defensive digging, grooming, approaching and touching the hand, aggression or other active behaviours were recorded. Statistical analysis has not revealed significant differences between the treatments groups, either (1) because the olfactory properties of the handling method does not affect these behaviours, (2) because the impact of handing itself makes the type of glove material (or lack of) irrelevant, or (3) because the number of animals used was insufficient to provide statistically significant results. Understanding whether and how husbandry odours affect mouse welfare will help to refine mouse husbandry, improving mouse welfare and experimental results

Spatial use of western gorillas at a new Kyoto City Zoo facility – an exhibition of arboreal gorillas

Masayuki Tanaka[1,2], Sato Maegaki[1], Fumio Itoh[3], Tomoko Sasaki[2,3] and Mitsunori Nagao[2,3]
[1]Kyoto University, Wildlife Research Center, 2-24 Tanaka-Sekidencho, Sakyo-ku, Kyoto, 606-8203 Kyoto, Japan, [2]Kyoto City Zoo, Center for Research and Education of Wildlife, Okazaki-Hosshojicho, Sakyo-ku, Kyoto, 606-8333 Kyoto, Japan, [3]Kyoto City Zoo, Animal Husbandry and Conservation Section, Okazaki-Hosshojicho, Sakyo-ku, Kyoto, 606-8333 Kyoto, Japan; tanaka.masayuki.wrc@gmail.com

Studies of wild western gorillas have shown they are highly arboreal, but gorillas in zoological exhibitions are often shown on the ground. Lack of wild western gorillas leads not to make three-dimensional structure. At the Kyoto City Zoo, a new facility for western gorillas called 'Gorilla House –Habitat in the Forest–' was opened in April, 2014. There are vertical posts, cross beams, and many ropes connecting the posts and beams outdoors. Natural trees have also been planted outside, and the gorillas are free to touch, pull, or break them. An adult male (14 yrs.), an adult female (28 yrs.), and an infant male (3 yrs.) were moved to this facility in April, 2014 to live together. They usually spent time outside from 9:30 to 15:30, and they were fed three times a day (9:30, 13:00, and 15:30). We observed their use of space to examine the effect of the new facility. The observation period was from June to December, 2014 (77 days in total). Using a tablet PC, an observer recorded the gorillas' locations and activity using a one-minute-scan sampling method. We observed the gorillas for 1 h in the morning (10:30-11:30) and afternoon (13:00-14:00). Observations showed that each gorilla made use of the area above three meters. The infant gorilla used the higher areas the most of the three, spending more than 50% of the observation time there. The female adult gorilla was seen there less frequently, but spent more than 10% of the observation time at above three meters. The results showed that the western gorillas frequently used the higher areas at our new facility. The infant and adult male gorillas were often seen on the ground, the latter frequently near the window where many visitors could see him. These results suggest that captive western gorillas spontaneously spent more time in higher area. The results also suggest that they did not avoid the ground, but selected any area of their liking.

The influence of puzzle feeders on the behaviour of domestic cats in a rescue shelter

Moira Harris
Harper Adams University, Department of Animal Production, Welfare and Veterinary Sciences, Newport, Shropshire TF10 8NB, United Kingdom; mharris@harper-adams.ac.uk

Rescue shelters can be extremely stressful for cats. Environmental enrichment may increase the range of normal behaviour and reduce abnormal behaviour; however, animals can habituate to enrichment items, limiting long-term benefits. Studies have examined the effects of olfactory and visual enrichment, but little research has been conducted on feed enrichment for domestic cats. This study examined the influence of two puzzle feeders on cat behaviour in a rescue shelter and whether the cats habituated to the devices. Twelve individual cats were selected and randomly assigned to one of two enrichment devices, either a ball (B) or a tower (T). These were filled with biscuits and placed in the cats' indoor enclosures for two hours, every day for five days. Behaviour was recorded using instantaneous sampling on days 1, 3 and 5. The study was repeated with the other enrichment device. Finally, cats were randomly assigned either B or T on day 1 and enrichments were alternated daily, with behaviour observed on days 4 and 5 to test for habituation. On day 1 cats spent more time interacting directly with T than B (one-way ANOVA; $P<0.001$) but by day 5 there was no difference between treatments. On day 5 both B and T cats showed less pacing behaviour compared to a control group (one-way ANOVA; $P<0.001$). No habituation to either B or T was exhibited during the study; levels of direct interaction did not differ between days 1 and 5. During the habituation study cats showed more interaction with other objects in their environment on day 5 than either B or T cats in the single enrichment trials (two-way ANOVA; $P=0.31$). Both devices reduced pacing, possibly indicating improved welfare. A longer study would be beneficial to fully evaluate the effects of time on cats' behavioural responses to enrichment.

Authors index